SMALL WELLS
MANUAL

Location, Design, Construction, Use and Maintenance

Ulric P. Gibson, B.Sc.Hons. (Civ. Eng.), M.S., M.I.C.E.,
  Division of Environmental Health, School of Public Health,
  University of Minnesota. Former Executive Engineer, Water
  Supply, Rural Areas, Ministry of Works & Hydraulics, Guyana.

Rexford D. Singer, B.S.C.E., M.S.,
  Division of Environmental Health, School of Public Health,
  University of Minnesota.

Agency for International Development

Books for Business
New York-Hong Kong

Small Wells Manual:
Location, Design, Construction, Use
and Maintenance

by
Ulric P. Gibson
Rexford D. Singer

for
Agency for International Development

ISBN: 0-89499-173-6

Reprinted from the 1969 edition

Books for Business
New York - Hong Kong
http://www.BusinessBooksInternational.com

SMALL WELLS
MANUAL

ACKNOWLEDGEMENTS

The authors wish to express their appreciation to the Health Service, Office of War on Hunger, United States Agency for International Development for making the publication of this manual possible. We are particularly indebted to the UOP-Johnson Division, Universal Oil Products Company, St. Paul, Minnesota for their advice and assistance in preparing the manuscript and for their contribution of valuable information and illustrations and to Mr. Arpad Rumy for the preparation of many of the illustrations. We also wish to express sincere gratitude to all persons who have offered comments, suggestions and assistance or who have given their time to critically review the manuscript.

In preparing this manual, an attempt has been made to bring together information and material from a variety of sources. We have endeavored to give proper credit for the direct use of material from these sources, and any omission of such credit is unintentional.

FOREWORD

It has been estimated that nearly two-thirds of the one and a half billion people living in the developing countries are without adequate supplies of safe water. The consequences of this deficiency are innumerable episodes of the debilitating and incapacitating enteric diseases which annually affect an estimated 500 million people and result in the deaths of as many as 10 million about half of whom are children.

Although there are many factors limiting the installation of small water systems, the lack of knowledge in regard to the availability of ground water and effective means of extracting it for use by rural communities is a major element. It is anticipated that this manual will make a major contribution toward filling this need by providing the man in the field, not necessarily an engineer or hydrologist, with the information needed to locate, construct and operate a small well which can provide good quality water in adequate quantities for small communities.

The Agency for International Development takes great pride in cooperating with the University of Minnesota in making this manual available.

Arthur H. Holloway
Sanitary Engineer,
Health Service, Office of War on Hunger
Agency for International Development

TABLE OF CONTENTS

CHAPTER 1

INTRODUCTION

PURPOSE

This manual is intended to serve as a basic introductory text book and to provide instruction and guidance to field personnel engaged in the construction, operation and maintenance of small diameter, relatively shallow wells used primarily for individual and small community water supplies.

It is aimed particularly at those persons who have had little or no experience in the subject. An attempt has been made to treat the subject matter as simply as possible in order that this manual may be of benefit not only to the engineer or other technically trained individual (inexperienced in this field) but also the individual home owner, farmer or non-technically trained community development officer. This manual should also prove useful in the training of water well drillers, providing the complementary background material for their field experience. The reader who is interested in pursuing the subject further, and with reference to larger and deeper wells, is referred to the list of references to be found at the end of this manual.

SCOPE

This manual covers the exploration and development of ground-water sources in *unconsolidated formations,* primarily for the provision of small potable water supplies. Its scope has been limited to the consideration of small tube wells up to *4 inches in diameter,* a maximum of approximately *100 feet in depth* and with yields of up to about *50 U.S. gallons per minute* (All references are to U.S. units. Conversion tables are to be found in Appendix B). The location, design, construction, maintenance and rehabilitation of such wells are among the various aspects discussed. The above limitation on well size (diameter) rules out the consideration of dug wells in favor of the much more efficient and easier to protect bored, driven, jetted or drilled tube wells. However, a method of converting existing dug wells to tube wells is discussed.

PUBLIC HEALTH AND RELATED FACTORS
Importance of Water Supplies

Water is, with the exception of air, the most important single substance to man's survival. Man, like all other forms of biological life, is extremely dependent upon water and can survive much longer without food than he can without water. The quantities of water directly required for the proper functioning of the body processes are relatively small but essential.

1

While man has always recognized the importance of water for his internal bodily needs, his recognition of its importance to health is a more recent development, dating back only a century or so. Since that time, much has been learned about the role of inadequate and contaminated water supplies in the spread of water-borne diseases. Among the first diseases recognized to be water borne were cholera and typhoid fever. Later, dysentery, gastro-enteritis and other diarrheal diseases were added to the list. More recently, water has also been shown to play an important role in the spread of certain viral diseases such as infectious hepatitis.

Water is involved in the spread of communicable diseases in essentially two ways. The first is the well known direct ingestion of the infectious agent when drinking contaminated water (e.g. dysentery, typhoid and other gastrointestinal diseases). The second is due to a lack of sufficient water for personal hygiene purposes. Inadequate quantities of water for the maintenance of personal hygiene and environmental sanitation have been shown to be major contributing factors in the spread of such diseases as yaws and typhus. Adequate supplies of water for personal hygiene also diminish the probability of transmitting some of the gastrointestinal diseases mentioned above. The latter type of interaction between water and the spread of disease has been recognized by various public health organizations in developing countries which have been trying to provide adequate quantities of water of reasonable, though not entirely satisfactory, quality.

Health problems related to the inadequacy of water supplies are universal but, generally, of greater magnitude and significance in the underdeveloped and developing nations. It has been estimated that about two-thirds of the population of the developing countries obtain their water from contaminated sources. The World Health Organization estimates that each year 500 million people suffer from diseases associated with unsafe water supplies. Due largely to poor water supplies, an estimated 5,000,000 infants die each year from diarrheal diseases.

In addition to the human consumption and health requirements, water is also needed for agricultural, industrial and other purposes. Though all of these needs are important, water for human consumption and sanitation is considered to be of greater social and economic importance since the health of the population influences all other activities.

Ground-Water's Importance

It can generally be said that ground water has played a much less imporatnt role in the solution of the world's water supply problems than its relative availability would indicate. Its out-of-sight location and the associated lack of knowledge with respect to its occurrence, movement and development have no doubt contributed greatly to this situation. The increasing acquisition and dissemination of knowledge pertaining to ground-water development will gradually allow the use of this source of water to approach its rightful degree of importance and usefulness.

More than 97 percent of the fresh water on our planet (excluding that in the polar ice-caps and glaciers) is to be found underground. While it is not practicable to extract all of this water because of economic and other reasons, the recoverable quantities would, no doubt, exceed the available supplies of fresh surface water found in rivers and lakes.

Ground-water sources also represent water that is essentially in storage while the water in rivers and lakes is generally in transit, being replaced several times a year. The available quantity of surface water at any given location is also more subject to seasonal fluctuations than is ground water. In many areas, the extraction of ground water can be continued long after droughts have completely depleted rivers. Ground-water sources are, therefore, more reliable sources of water in many instances.

As will be seen in Chapter 2, ground waters are usually of much better quality than surface waters, due to the benefits of percolation through the ground. Oftener than not, ground water is also more readily available where needed, requiring less transportation and, generally, costing less to develop. Greater emphasis should, therefore, be placed on the development and use of the very extensive ground-water sources to be found throughout the world.

Need for Proper Development and Management of Ground-Water Resources

While some ground-water reservoirs are being replenished year after year by infiltration from precipitation, rivers, canals and so on, others are being replenished to much lesser degrees or not at all. Extraction of water from these latter reservoirs results in the continued depletion or mining of the water.

Ground water also often seeps into streams, thus providing the low flow (base flow) that is sustained through the driest period of the year. Conversely, if the surface water levels in streams are higher than those in ground-water reservoirs, then seepage takes place in the opposite direction, from the streams into the ground-water reservoirs. Uncontrolled use of ground water can, therefore, affect the levels of streams and lakes and consequently the uses to which they are normally put.

Ground-water development presents special problems. The lack of solutions to these problems have, in the past, contributed to the mystery that surrounded ground-water development and the limited use to which ground water has been put. The proper development and management of ground-water resources requires a knowledge of the extent of storage, the rates of discharge from and recharge to underground reservoirs, and the use of economical means of extraction. It may be necessary to devise artificial means of recharging these reservoirs where no natural sources exist or to supplement the natural recharge. Research has, in recent years, considerably increased our knowledge of the processes involved in the origin and movement of ground water and has provided us with better methods of development and conservation of ground-water supplies. Evidence of this increased knowledge is to be found in the greater emphasis being placed on ground-water development.

CHAPTER 2

ORIGIN, OCCURRENCE AND MOVEMENT
OF GROUND WATER

An understanding of the processes and factors affecting the origin, occurrence, and movement of ground water is essential to the proper development and use of ground-water resources. Of importance in determining a satisfactory rate of extraction and suitable uses of the water are a knowledge of the quantity of water present, its origin, the direction and rate of movement to its point of discharge, the discharge rate and the rate at which it is being replenished, and the quality of the water. These points are considered in this chapter in as simplified and limited a form as the aims and scope of this manual permit.

THE HYDROLOGIC CYCLE

The hydrologic cycle is the name given to the circulation of water in its liquid, vapor, or solid state from the oceans to the air, air to land, over the land surface or underground, and back to the oceans (Fig. 2.1).

Evaporation, taking place at the water surface of oceans and other open bodies of water, results in the transfer of water vapor to the atmosphere. Under certain conditions, this water vapor condenses to form clouds which subsequently release their moisture as precipitation in the form of rain, hail, sleet, or snow. Precipitation may occur over the oceans returning some of the water directly to them or over land to which winds have previously transported the moisture-laden air and clouds. Part of the rain falling to the earth evaporates with immediate return of moisture to the atmosphere. Of the remainder, some, upon reaching the ground surface, wets it and runs off into surface streams finally discharging in the ocean while another part infiltrates into the ground and then percolates to the ground-water flow through which it later reaches the ocean. Evaporation returns some of the water from the wet land surface to the atmosphere while plants extract some of that portion in the soil through their roots and, by a process known as transpiration, return it through their leaves to the atmosphere.

SUBSURFACE DISTRIBUTION OF WATER

Subsurface water found in the interstices or pores of rocks may be divided into two main zones (Fig. 2.2). These are the *zone of aeration* and the *zone of saturation.*

Zone of Aeration

The zone of aeration extends from the land surface to the level at which all of the pores or open spaces in the earth's materials are completely filled or

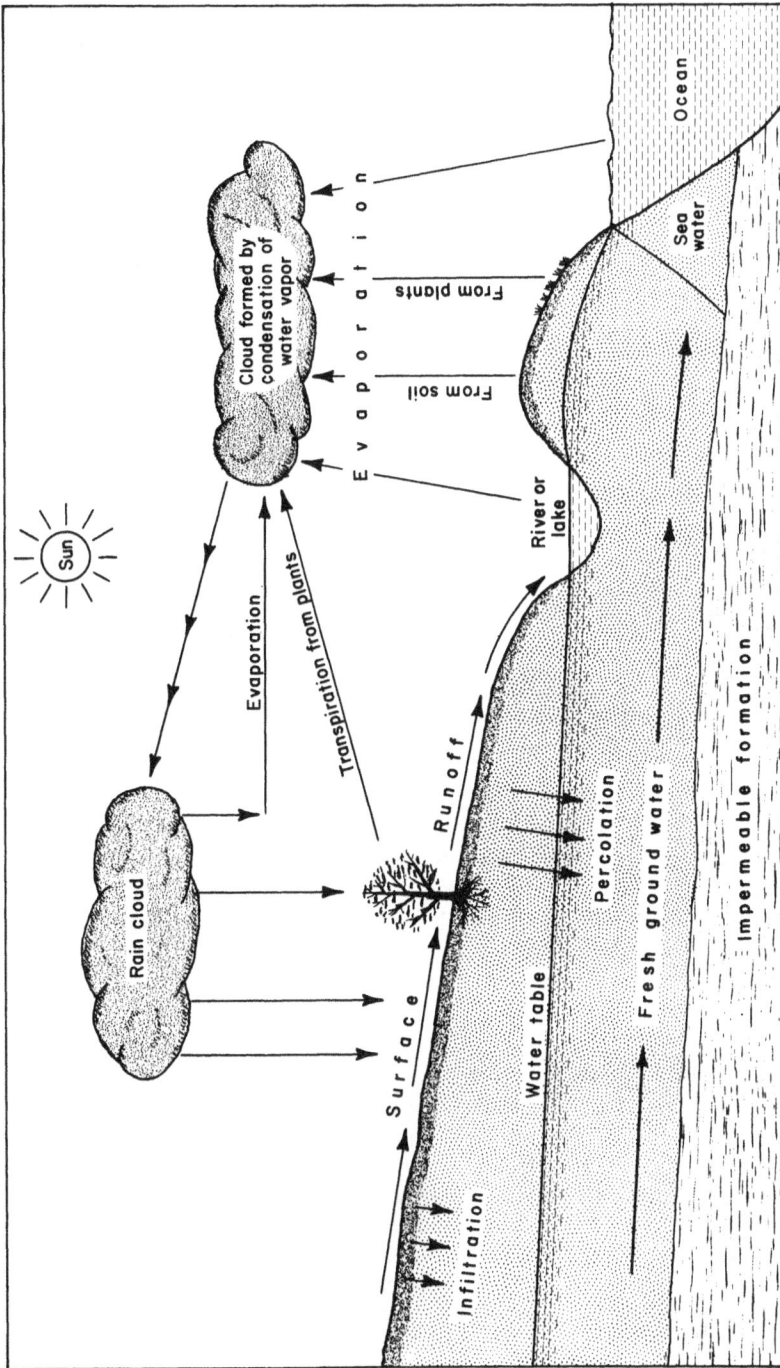

Fig. 2.1 THE HYDROLOGIC CYCLE.

```
          Land Surface

       Belt of Soil Water

                                    Zone of Aeration
       Intermediate Belt


       Capillary Fringe
         Water Table
       ─ ─ ─ ─ ─ ─ ─ ─ ─ ─
       ─ ─ ─ ─ ─ ─ ─ ─ ─ ─
       ─ ─ ─ ─ ─ ─ ─ ─ ─ ─   Zone of Saturation
       ─ ─ ─ ─ ─ ─
       Ground Water
```

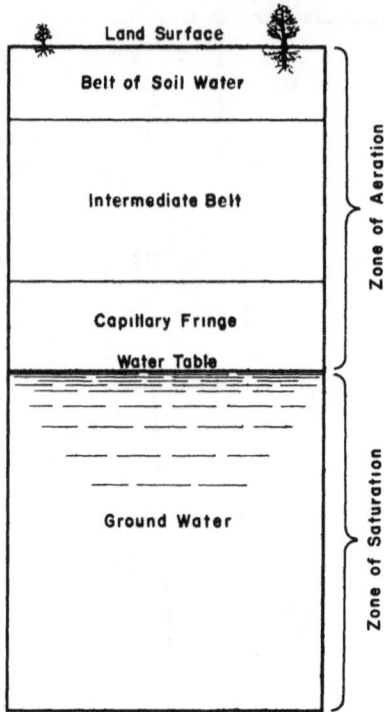

Fig. 2.2 DIVISIONS OF SUBSUR-FACE WATER.

saturated with water. A mixture of air and water is to be found in the pores in this zone and hence its name. It may be sub-divided into three belts. These are (1) the belt of soil water, (2) the intermediate belt and (3) the capillary fringe.

The *belt of soil water* lies immediately below the surface and is that region from which plants extract, by their roots, the moisture necessary for growth. The thickness of the belt differs greatly with the type of soil and vegetation, ranging from a few feet in grass-lands and field crop areas to several feet in forests and lands supporting deep-rooted plants.

The *capillary fringe* occupies the bottom portion of the zone of aeration and lies immediately above the zone of saturation. Its name comes from the fact that the water in this belt is suspended by capillary forces similar to those which cause water to rise in a narrow or capillary tube above the level of the water in a larger vessel into which the tube has been placed upright. The narrower the tube or the pores, the higher the water rises. Hence, the thickness of the belt depends upon the texture of the rock or soil and may be practically zero where the pores are large.

The *intermediate belt* lies between the belt of soil water and the capillary fringe. Most of its water reaches it by gravity drainage downward through the belt of soil water. The water in this belt is called intermediate (vadose) water.

Zone of Saturation

Immediately below the zone of aeration lies the zone of saturation in which the pores are completely filled or saturated with water. The water in the zone of saturation is known as *ground water* and is the only form of subsurface water that will flow readily into a well. The object of well construction is to penetrate the earth into this zone with a tube, the bottom section of which has openings which are sized such as to permit the inflow of water from the zone of saturation but to exclude its rock particles. Formations which contain ground water and will readily yield it to wells are called *aquifers*.

GEOLOGIC FORMATIONS AS AQUIFERS

For convenience, geologists describe all earth materials as *rocks*. Rocks may be of the *consolidated* type (held firmly together by compaction, cementation and other processes) such as granite, sandstone and limestone or *unconsolidated* type (loose materials) such as clay, sand and gravel. The terms *hard* and *soft* are also used to describe consolidated and unconsolidated rocks respectively.

Aquifers may be composed of consolidated or unconsolidated rocks. The rock materials must be sufficiently porous (contain a reasonably high proportion of pores or other openings to solid material) and be sufficiently permeable (the openings must be interconnected to permit the travel of water through them).

Rock Classification

Rocks may be classified with respect to their origin into the three main categories of sedimentary rocks, igneous rocks, and metamorphic rocks.

Sedimentary rocks are the deposits of material derived from the weathering and erosion of other rocks. Though constituting only about 5 percent of the earth's crust they contain an estimated 95 percent of the available ground water.

Sedimentary rocks may be consolidated or unconsolidated depending upon a number of factors such as the type of parent rock, mode of weathering, means of transport, mode of deposition, and the extent to which packing, compaction, and cementation have taken place. Harder rocks generally produce sediments of coarser texture than softer ones. Weathering by mechanical disintegration (e.g. rock fracture due to temperature variations) produces coarser sediments than those produced by chemical decomposition. Deposition in water provides more sorting and better packing of materials than does deposition directly onto land. Chemical constituents in the parent rocks and the environment are responsible for the cementation of unconsolidated rocks into hard, consolidated ones. These factors also influence the water-bearing capacity of sedimentary rocks. Disintegrated shale sediments are usually fine-grained and make poor aquifers while sediments derived from granite or other crystalline rocks usually form good sand and gravel aquifers, particularly when considerable water transportation has resulted in well-rounded and sorted particles.

Sand, gravel, and mixtures of sand and gravel are among the unconsolidated sedimentary rocks that form aquifers. Granular and unconsolidated, they vary in particle size and in the degree of sorting and rounding of the particles. Consequently, their water-yielding capabilities vary considerably. However, they consitute the best water-bearing formations. They are widely distributed throughout the world and produce very significant proportions of the water used in many countries.

Other unconsolidated sedimentary aquifers include marine deposits, alluvial or stream deposits (including deltaic deposits and alluvial fans), glacial drifts and wind-blown deposits such as dune sand and loess (very fine silty deposits). Great variations in the water-yielding capabilities of these formations can also be expected. For example, the yield from wells in sand dunes and loess may be limited by both the fineness of the material and the limited areal extent and thickness of the deposits.

Limestone, essentially calcium carbonate, and dolomite or calcium-magnesium carbonate are examples of consolidated sedimentary rocks known to function as aquifers. Fractures and crevices caused by earth movement, and later enlarged into solution channels by ground-water flow through them, form the connected openings through which flow takes place (Fig. 2.3). Flows may be considerable where solution channels have developed.

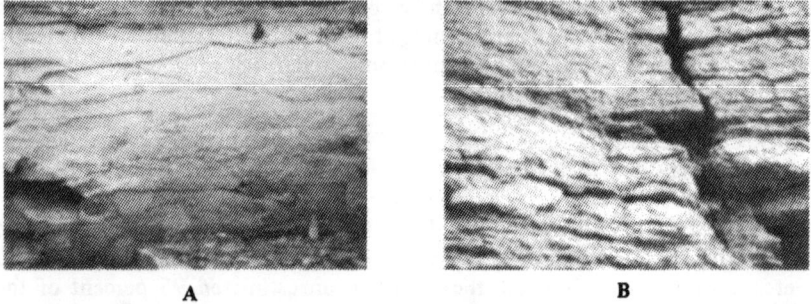

A   B

Fig. 2.3   A. FRACTURES IN DENSE LIMESTONE THROUGH WHICH FLOW MAY OCCUR.
B. SOLUTION CHANNELS IN LIMESTONE CAUSED BY GROUND-WATER FLOW THROUGH FRACTURES.

Sandstone, usually formed by compaction of sand deposited by rivers near existing sea shores, is another form of consolidated sedimentary rock that performs as an aquifer. The cementing agents are responsible for the wide range of colors seen in sandstones. The water-yielding capabilities of sandstones vary with the degree of cementation and fracturing.

Shales and other similar compacted and cemented clays, such as mudstone or siltstone, are usually not considered to be aquifers but have been known to yield small quantities of water to wells in localized areas where earth movements have substantially fractured such formations.

Igneous rocks are those resulting from the cooling and solidification of hot, molten materials called magma which originate at great depths within the earth. When solidification takes place at considerable depth, the rocks are referred to as *intrusive* or *plutonic* while those solidifying at or near the ground surface are called *extrusive* or *volcanic*.

Plutonic rocks such as granite are usually coarse-textured and non-porous and are not considered to be aquifers. However, water has occasionally been found in crevices and fractures of the upper, weathered portions of such rocks.

Volcanic rocks, because of the relatively rapid cooling taking place at the surface, are usually fine-textured and glassy in appearance. Basalt or trap rock, one of the chief rocks of this type, can be highly porous and permeable as a result of interconnected openings called vesicles formed by the development of gas bubbles as the lava (magma flowing at or near the surface) cools. Basaltic aquifers may also contain water in crevices and broken up or brecciated tops and bottoms of successive layers.

8

Fragmental materials discharged by volcanos, such as ash and cinders, have been known to form excellent aquifers where particle sizes are sufficiently large. Their water-yielding capabilities vary considerably, depending on the complexity of stratification, the range of particle sizes, and shape of the particles. Examples of excellent aquifers of this type are to be found in Central America.

Metamorphic rock is the name given to rocks of all types, igneous or sedimentary, which have been altered by heat and pressure. Examples of these are quartzite or metamorphosed sandstone, slate and mica schist from shale, and gneiss from granite. Generally, these form poor aquifers with water obtained only from cracks and fractures. Marble, a metamorphosed limestone, can be a good aquifer when fractured and containing solution channels.

With the above description of the three main rock types, it should now be easier to understand why an estimated 95 percent of the available ground water is to be found in sedimentary rocks which constitute only about 5 percent of the earth's crust. The wells described in this manual will be those constructed in unconsolidated sedimentary rocks which are undoubtedly the most important sources of water for small community water supply systems.

Role of Geologic Processes in Aquifer Formation

Geologic processes are continually, though slowly, altering rocks and rock formations. So slowly are these changes taking place that they are hardly perceptible to the human eye and only barely measurable by the most sensitive instruments now available. Undoubtedly, however, mountains are being up-lifted and lowered, valleys filled or deepened and new ones created, sea shores advancing and retreating, and aquifers created and destroyed. These changes are more obvious when referred to a geologic timetable with units measured in thousands and millions of years and to which reference can be made in almost any book on geology.

Geologically old as well as young rocks may form aquifers but generally the younger ones which have been subjected to less compression and cementation are the better producers. Geologic processes determine the shape, extent, and hydraulic or flow characteristics of aquifers. Aquifers in sedimentary rock formations for example vary considerably depending upon whether the sediments are terrestrial or marine in nature.

Terrestrial sediments, or materials deposited on land, include stream, lake, glacial, and wind-blown deposits. With but few exceptions they are usually of limited extent and discontinuous, much more so than are marine deposits. Texture variations both laterally and vertically are characteristic of these formations.

Alluvial or *stream deposits* are generally long and narrow. Usually subsurface, or below the valley floor, they may also be in the form of terraces indicating the existence of higher stream beds in the geologic past. The material in such aquifers may range in size from fine sand to gravel and boulders. Abandoned stream courses and their deposits are sometimes buried under wind-borne or glacial deposits with no visible evidence of their existence. Where a rapidly flowing stream such as a mountain stream encounters a rapid reduction of slope, the decrease in velocity causes a deposition of large aprons of material known as alluvial fans. These sediments range from coarser to finer material as one proceeds away from the mountains.

Glacial deposits found in North Central U.S.A., Southern Canada, and Northern Europe and Asia may be extensive where they result from continental glaciers as compared to the more localized deposits of mountain glaciers. These deposits vary in shape and thickness and exhibit a lack of interconnection because of the clay and silt accumulations within the sand, gravel and boulders. Outwash deposits swept out of the melting glacier by melt-water streams are granular in nature and similar to alluvial sands. The swifter melt-water streams produce the best glacial drift aquifers.

Lake deposits are generally fine-textured, granular material deposited in quiet water. They vary considerably in thickness, extent, and shape and make good aquifers only when they are of substantial thickness.

GROUND-WATER FLOW AND ELEMENTARY WELL HYDRAULICS
Types of Aquifers

Ground-water aquifers may be classified as either *water-table* or *artesian aquifers.*

A *water-table aquifer* is one which is not confined by an upper impermeable layer. Hence, it is also called an *unconfined* aquifer. Water in these aquifers is virtually at atmospheric pressure and the upper surface of the zone of saturation is called the *water table* (Fig. 2.2). The water table marks the highest level to which water will rise in a well constructed in a water-table aquifer. The upper aquifer in Fig. 2.4 is an example of a water-table aquifer.

An *artesian aquifer* is one in which the water is confined under a pressure greater than atmospheric by an overlying, relatively impermeable layer. Hence, such aquifers are also called *confined* or *pressure* aquifers. The name artesian owes its origin to Artois, the northernmost province of France, where the first deep wells to tap confined aquifers were known to have been drilled. Unlike water-table aquifers, water in artesian aquifers will rise in wells to levels above the bottom of the upper confining layer. This is because of the pressure created by that confining layer and is the distinguishing feature between the two types of aquifers.

The imaginary surface to which water will rise in wells located throughout an artesian aquifer is called the *piezometric surface*. This surface may be either above or below the ground surface at different parts of the same aquifer as is shown in Fig. 2.4. Where the piezometric surface lies above the ground surface, a well tapping the aquifer will flow at ground level and is referred to as a *flowing artesian* well. Where the piezometric surface lies below the ground surface, a *non-flowing artesian well* results and some means of lifting water, such as a pump, must be provided to obtain water from the well. It is worthy of note here that the earlier usage of the term artesian well referred only to the flowing type while current usage includes both flowing and non-flowing wells, provided the water level in the well rises above the bottom of the confining layer or the top of the aquifer.

10

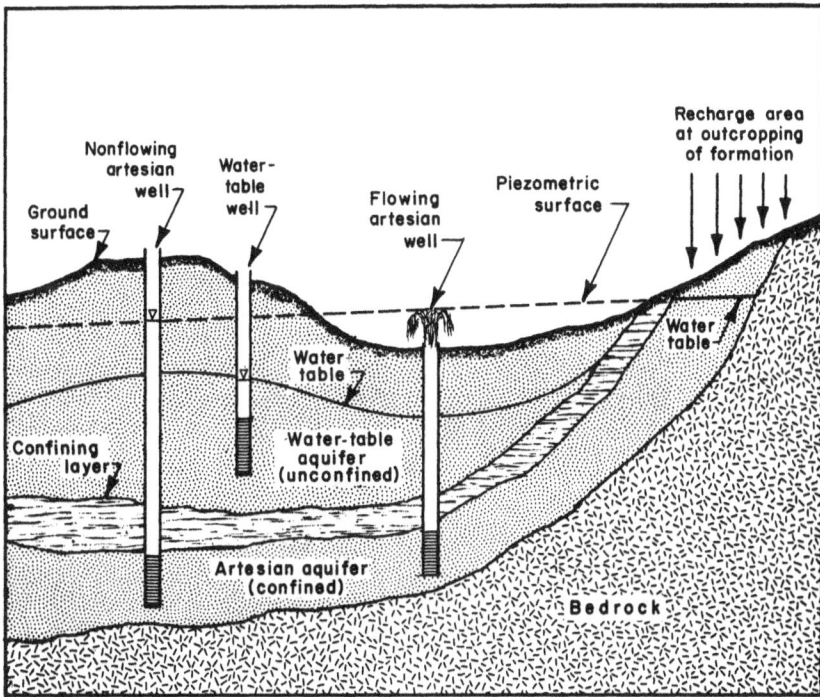

Fig. 2.4 TYPES OF AQUIFERS.

Water usually enters an artesian aquifer in an area where it rises to the ground surface and is exposed (Fig. 2.4). Such an exposed area is called a *recharge area* and the aquifer in that area, being unconfined, would be of the water-table type. Artesian aquifers may also receive water underground from leakage through the confining layers and at intersections with other aquifers, the recharge areas of which are at ground level.

Aquifer Functions

The openings and pores in a water-bearing formation may be considered as a network of interconnected pipes through which water flows at very slow rates, seldom more than a few feet per day, from areas of recharge to areas of discharge. This network of pipes, therefore, serves to provide both *storage* and flow or *conduit* functions in an aquifer.

Storage function: Related to the storage function of an aquifer are two important properties known as *porosity* and *specific yield.*

The *porosity* of a water-bearing formation is that percentage of the total volume of the formation which consists of openings or pores. For example, the porosity of one cubic foot of sand which contains 0.25 cubic foot of open spaces is 25 percent. It is therefore evident that porosity is an index of the amount of ground water that can be stored in a saturated formation.

The amount of water yielded by, or that may be taken from, a saturated formation is less than that which it holds and is, therefore, not represented by

11

the porosity. This quantity is related to the property known as the *specific yield* and defined as the volume of water released from a unit volume of the aquifer material when allowed to drain freely by gravity (Fig. 2.5). The remaining volume of water not removed by gravity drainage is held by capillary forces such as found in the capillary fringe and by other forces of attraction. It is called the *specific retention* and, like the specific yield, may be expressed as a decimal fraction or percentage. As defined, porosity is therefore equal to the sum of the specific yield and the specific retention. An aquifer with a porosity of 0.25 or 25 percent and a specific yield of 0.10 or 10 percent would, therefore, have a specific retention of 0.15 or 15 percent. One million cubic feet of such an aquifer would contain 250,000 cubic feet of water of which 100,000 cubic feet would be yielded by gravity drainage.

Conduit function: The property of an aquifer related to its conduit function is known as the *permeability.*

Permeability is a measure of the capacity of an aquifer to transmit water. It is related to the pressure difference and velocity of flow between two points under laminar or non-turbulent conditions by the following equation known as Darcy's Law (after Henry Darcy, the French engineer who developed it).

$$V = \frac{P(h_1 - h_2)}{\ell} \tag{2.1}$$

where V   is the velocity of flow in feet per day,

h_1 is the pressure at the point of entrance to the section of conduit under consideration in feet of water,

h_2 is the pressure at the point of exit of the same section in feet of water,

ℓ is the length of the section of conduit in feet, and

P   is a constant known as the coefficient of permeability but often referred to simply as the permeability.

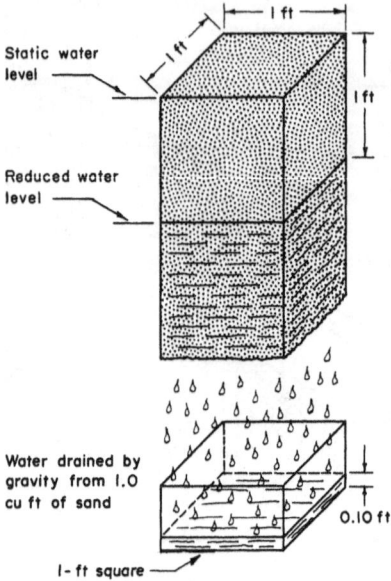

Fig. 2.5 VISUAL REPRESENTATION OF SPECIFIC YIELD. ITS VALUE HERE IS 0.10 CU FT PER CU FT OF AQUIFER MATERIAL.

Equation (2.1) may be modified to read

$$V = PI \qquad (2.2)$$

where $I = \dfrac{h_1 - h_2}{\ell}$, and is called the *hydraulic gradient*.

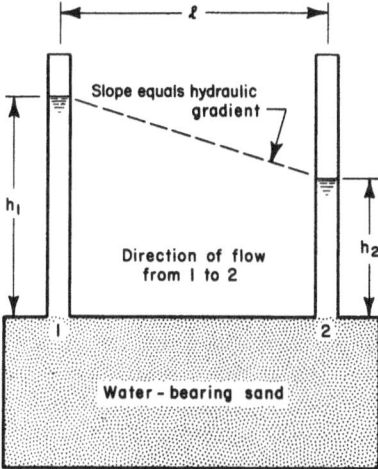

Fig 2.6 SECTION THROUGH WATER-BEARING SAND SHOWING THE PRESSURE DIFFERENCE $(h_1 - h_2)$ CAUSING FLOW BETWEEN POINTS 1 AND 2. THE HYDRAULIC GRADIENT IS EQUAL TO THE PRESSURE DIFFERENCE DIVIDED BY THE DISTANCE, ℓ, BETWEEN THE POINTS.

The quantity of flow per unit of time through a given cross-sectional area may be obtained from equation (2.2) by multiplying the velocity of flow by that area. Thus,

$$Q = AV = PIA \qquad (2.3)$$

where Q is the quanity of flow per unit of time

and A is the cross-sectional area.

Based on equation (2.3) the *coefficient of permeability* may, therefore, be defined as the quantity of water that will flow through a unit cross-sectional area of porous material in unit time under a hydraulic gradient of unity (or $I = 1.0$) at a specified temperature, usually taken as 60°F. In ground-water problems, Q is usually expressed in gallons per day (gpd), A in square feet (sq ft) and P, therefore, in gallons per day per square foot (gpd/sq ft). The coefficient of permeability can also be expressed in the metric system using units of liters per day per square meter under a hydraulic gradient of unity and at a temperature of 15.5°C.

It is important to note that Darcy's Law in the form shown in equation (2.3) states that the quantity of water flowing under laminar or non-turbulent conditions varies in direct proportion to the hydraulic gradient and, therefore, the pressure difference $(h_1 - h_2)$ causing the flow. This means that doubling the pressure difference will result in doubling the flow through the same cross-sectional area. By definition, the hydraulic gradient is seen to be equivalent to the slope of the water table for a water-table aquifer or of the piezometric surface for an artesian aquifer.

Considering a vertical cross-section of an aquifer of unit width and having a total thickness, m, a hydraulic gradient, I, and an average coefficient of

13

permeability, P, we see from equation (2.3) that the rate of flow, q, through this cross section is given by

$$q = PmI \qquad (2.4)$$

The product Pm of equation (2.4) is termed the *coefficient of transmissibility* or transmissivity, T, of the aquifer. By further considering that the total width of the aquifer is W, then the rate of flow, Q, through a vertical cross-section of the aquifer is given by

$$Q = qW = TIW \qquad (2.5)$$

The *coefficient of transmissibility* is, therefore, defined as the rate of flow through a vertical cross-section of an aquifer of unit width and whose height is the total thickness of the aquifer when the hydraulic gradient is unity. It is expressed in gallons per day per foot (gpd/ft) and is equivalent to the product of the coefficient of permeability and the thickness of the aquifer.

Factors Affecting Permeability

Porosity is an important factor affecting the permeability and, therefore, the capacity of an aquifer for yielding water. This is clearly evident since an aquifer can yield only a portion of the water that it contains and the higher the porosity, the greater is the volume of water that can be stored. Porosity must, however, be considered together with other related factors such as *particle size, arrangement* and *distribution, continuity of pores,* and formation *stratification.*

The volume of voids or pores associated with the closest packing of uniformly-sized spheres (Fig. 2.7) will represent the same percentage of the total volume (solids and voids) whether the spheres were all of tennis ball size or all 1/1000 inch in diameter. However, the smaller pores between the latter spheres would offer greater resistance to flow and, therefore, cause a decrease

Fig. 2.7 UNIFORMLY SIZED SPHERES PACKED IN RHOMBOHEDRAL ARRAY.

Fig. 2.8 UNIFORMLY SIZED SPHERES PACKED IN CUBICAL ARRAY.

in permeability even though the porosity is the same.

The packing of the spheres displayed in Fig. 2.7 is referred to as the rhombohedral packing. The porosity for such a packing can be shown to be 0.26 or 26 percent. The spheres may also assume a cubical array as shown in Fig. 2.8 for which the porosity is 0.476 or 47.6 percent. These porosities apply only to perfectly spherical particles and are included here to give the order of magnitude of the porosities that naturally occurring uniform sands and gravels may approach. A loose uniform sand may, for example, have a porosity of 46 percent. Clays, on the other hand, exhibit much higher porosities ranging from about 37 percent for stiff glacial clays to as high as 84 percent for soft bentonite clays.

Consideration of the effects of particle size and arrangement on permeability would be incomplete without simultaneously considering the effect of *particle distribution* or grading. A uniformly graded sand, that is, one in which all the particles are about the same size, will have a higher porosity and permeability than a less uniform sand and gravel mixture. This is so because the finer sand fills the openings between the gravel particles resulting in a more compact arrangement and less pore volume (Fig. 2.9). Here, then, is an example of a finer material having a higher permeability than a coarser one due to the modifying effect of particle distribution.

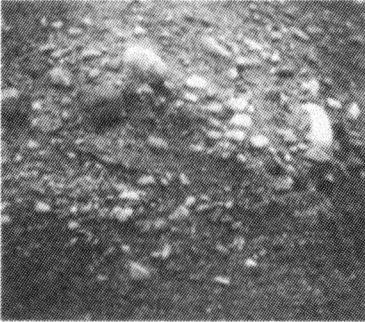

Fig. 2.9   NON-UNIFORM MIXTURE OF SAND AND GRAVEL WITH LOW POROSITY AND PERMEABILITY.

Flow cannot take place through porous material unless the passages in the material are interconnected, that is to say, there is *continuity of the pores*. Since permeability is a measure of the rate of flow under stated conditions through porous material, then a reduction in the continuity of the pores would result in a reduction in the permeability of the material. Such a reduction could be caused by silt, clay, or other cementing materials partially or completely filling the pores in a sand, thus making it almost impervious.

An aquifer is said to be *stratified* when it consists of different layers of fine sand, coarse sand, or sand and gravel. Most aquifers are stratified. While some strata contain silt and clay, others are relatively free from these cementing materials and are said to be clean. Where stratification is such that even a thin layer of clay separates two layers of clean sand, this results in the cutting off of the vertical movement of water between the sands. Permeability may also vary from layer to layer in a stratified aquifer.

A brief discussion on the measurement of permeability is to be found in Appendix A.

15

Flow Toward Wells

Converging flow: When a well is at rest, that is, when there is no flow taking place from it, the water pressure within the well is the same as that in the formation outside the well. The level at which water stands within the well is known as the *static water level.* This level coincides with the water table for a water-table aquifer or the piezometric surface for an artesian aquifer. Should the pressure be lowered within the well, by a pump for example, then the greater pressure in the aquifer on the outside of the well would force water into the well and flow thereby results. This lowering of the pressure within the well is also accompanied by a lowering of the water level in and around the well. Water flows through the aquifer to the well from all directions in what is known as *converging flow.* This flow may be considered to take place through successive cylindrical sections which become smaller and smaller as the well is approached (Fig. 2.10). This means that the area across which the flow takes place also becomes successively smaller as the well is approached. With the same quantity of water flowing across these sections, it follows from equation (2.3) that the velocity increases as the area becomes smaller.

Darcy's Law, equation (2.2), tells us that the hydraulic gradient varies in direct proportion to the velocity. The increasing velocity towards the well is, therefore, accompanied by an increasing hydraulic gradient. Stated in other terms, the water surface or the piezometric surface develops an increasingly steeper slope toward the well. In an aquifer of uniform shape and texture, the depression of the water table or piezometric surface in the vicinity of a pumped or freely flowing well takes the form of an inverted cone. This cone, known as the *cone of depression*

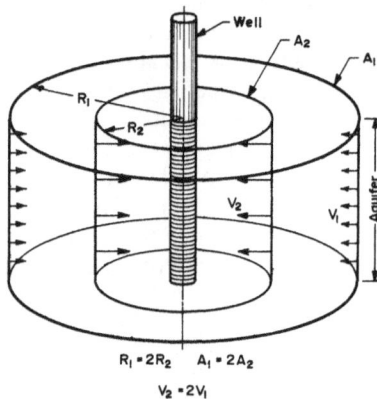

Fig. 2.10 F L O W CONVERGES TOWARD A WELL, PASSING THROUGH IMAGINARY CYLINDRICAL SURFACES THAT ARE SUCCESSIVELY SMALLER AS THE WELL IS APPROACHED.

(Fig. 2.11), has its apex at the water level in the well during pumping, and its base at the static water level. The water level in the well during pumping is known as the *pumping water level.* The difference in levels between the static water level and the surface of the cone of depression is known as the *drawdown.* Drawdown, therefore, increases from zero at the outer limits of the cone of depression to a maximum in the pumped well. The *radius of influence* is the distance from the center of the well to the outer limit of the cone of depression.

Fig. 2.12 shows how the transmissibility of an aquifer affects the shape of the cone of depression. The cone is deep, with steep sides, a large drawdown,

16

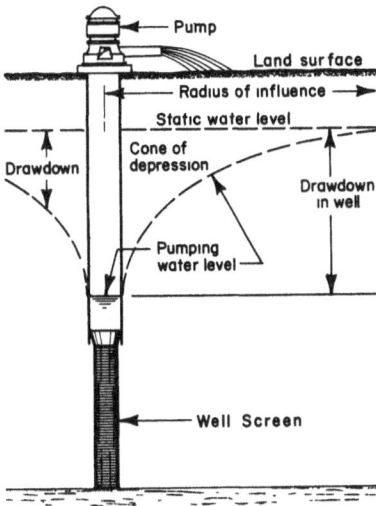

Fig. 2.11 CONE OF DEPRESSION IN VICINITY OF PUMPED WELL.

and a small radius of influence when the aquifer transmissibility is low. With a high transmissibility, the cone is wide and shallow, the drawdown being small, and the radius of influence large.

Recharge and boundary effects: When pumping commences at a well, the initial quantity of water discharged comes from the aquifer storage immediately surrounding the well. The cone of depression is then small. As pumping continues, the cone expands to meet the increasing demand for water from the aquifer storage. The radius of influence increases and, with it, the drawdown in the well in order to provide the additional pressure head required to move the water through correspondingly greater distances. If the rate of pumping is kept constant, then the rate of

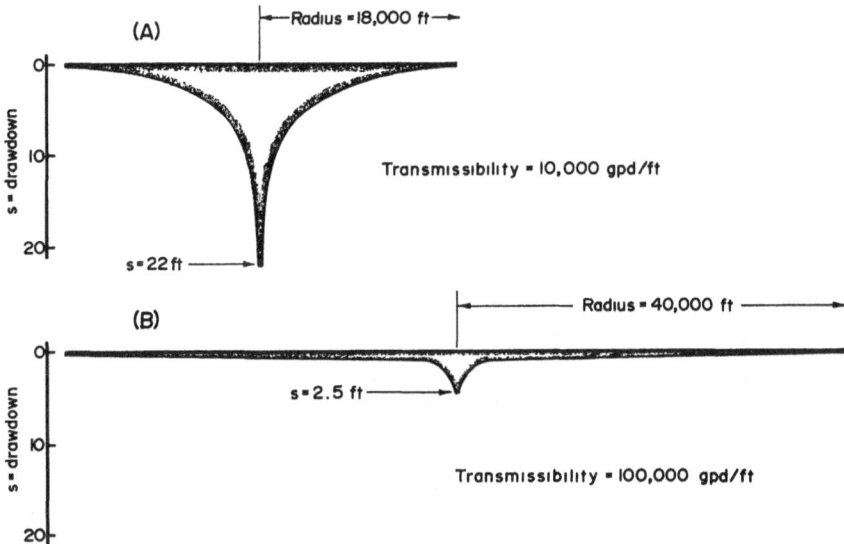

Fig. 2.12 EFFECT OF DIFFERING COEFFICIENTS OF TRANSMISSIBILITY UPON THE SHAPE, DEPTH AND EXTENT OF THE CONE OF DEPRESSION, PUMPING RATE AND OTHER FACTORS BEING THE SAME IN BOTH CASES.

17

expansion and deepening of the cone of depression decreases with time. This is illustrated in Fig. 2.13 where C_1, C_2 and C_3 represent cones of depression at hourly intervals. The hourly increases in radius of influence, R, and drawdown, s, become smaller and smaller until the aquifer supplies a quantity of water equal to the pumping rate. The cone no longer expands or deepens and *equilibrium* is said to have been reached. This state may occur in any one or more of the following situations.

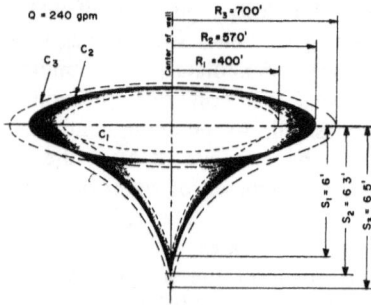

Fig. 2.13 CHANGES IN RADIUS AND DEPTH OF CONE OF DEPRESSION AFTER EQUAL INTERVALS OF TIME, ASSUMING CONSTANT PUMPING RATE.

1. The cone enlarges until it intercepts enough of the natural discharge from the aquifer to equal the pumping rate.
2. The cone intercepts a body of surface water from which water enters the aquifer at a rate equivalent to the pumping rate.
3. Recharge equal to the pumping rate is received from precipitation and vertical infiltration within the radius of influence.
4. Recharge equal to the pumping rate is obtained by leakage through adjacent formations.

Where the recharge rate is the same from all directions around the well the cone remains symmetrical (Fig. 2.12). If, however, it occurs mainly from one direction, as may be the case with a surface stream, then the surface of the cone is higher in the direction from which the recharge takes place than in other directions (Fig. 2.14). Conversely, the surface of the cone is relatively depressed in the direction of an impermeable boundary intercepted by it (Fig. 2.15). No recharge is obtained from such a boundary while that received from other directions maintains the higher levels in those directions. Recharge areas to aquifers, such as surface streams are, therefore, often referred to as positive boundaries while impermeable areas are known as negative boundaries.

Multiple well systems: Under some conditions the construction of a single large well may be either impractical or very costly while the installation of a group of small wells may be readily and economically accomplished. Factors such as the inaccessibility of the area to the heavy equipment required for drilling the large well and the high cost of transporting large diameter pipes to the site may be among the important considerations in a situation such as this. Small wells can be grouped in a proper pattern to give the equivalent performance of a much larger single well.

The grouping of wells, however, presents problems due to interference among them when operating simultaneously. *Interference* between two or more wells occurs when their cones of depression overlap, thus reducing the

18

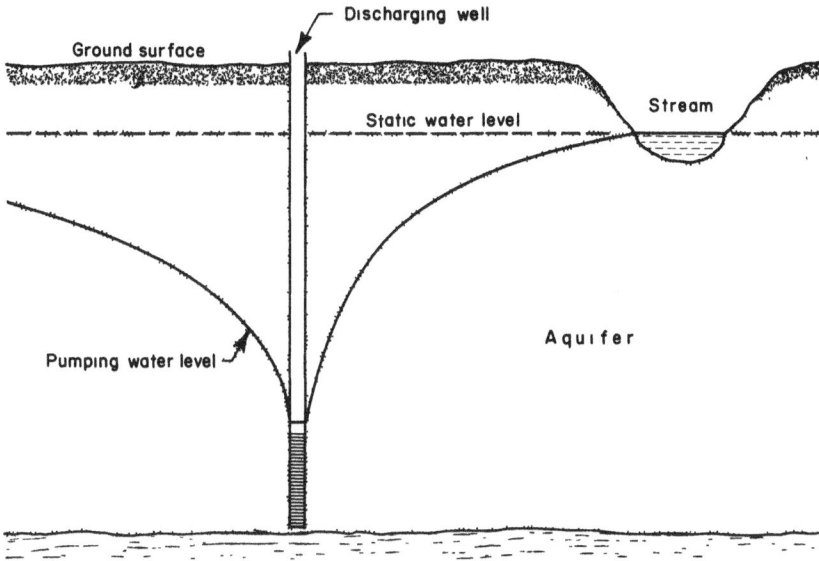

Fig. 2.14 SYMMETRY OF CONE OF DEPRESSION AFFECTED BY RECHARGE FROM STREAM.

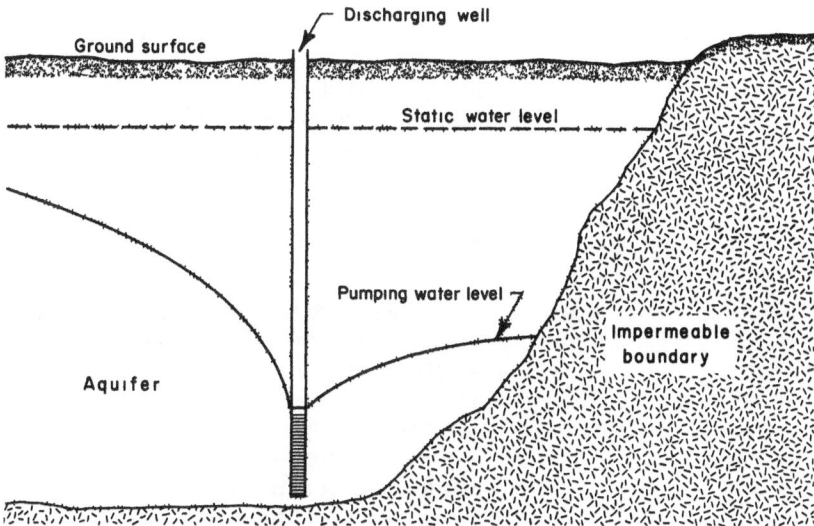

Fig. 2.15 CONE OF DEPRESSION IN VICINITY OF IMPERMEABLE BOUNDARY.

yield of the individual wells (Fig. 2.16). The drawdown at any point on the composite cone of depression is equal to the sum of the drawdowns at that point due to each of the wells being pumped separately. In particular, the drawdown for a specific discharge in a well affected by interference is greater than the unaffected value by the amount of drawdown at that well contributed by the interfering wells. In other words, the discharge per unit of drawdown commonly called the *specific capacity* of the well is reduced. This means that pumping must take place from a greater depth in the well, at a greater cost, to produce the same quantity of water from the well if it were not subject to interference.

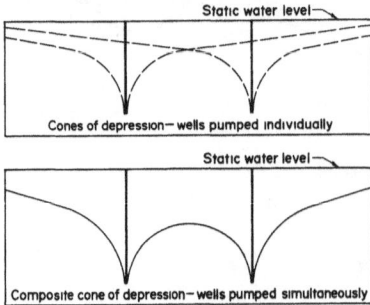

Fig. 2.16 INTERFERENCE BETWEEN ADJACENT WELLS TAPPING THE SAME AQUIFER.

Ideally, the solution would be to space the wells far enough apart to avoid the mutual interference of one on the other. Very often this is not practical for economic reasons and the wells are spaced far enough apart, not to eliminate interference, but to reduce it to acceptable proportions. For wells used for water supply purposes, spacings of 25 to 50 feet between wells have been found to be satisfactory. Spacings may be less in fine sand formations, in thin aquifers or when the drawdown is not likely to exceed 5 feet. Greater spacings may be used where the depth and thickness of the aquifer are such as to permit the use of screen lengths in excess of 10 feet.

There are many patterns which may be used when grouping wells (Fig. 2.17). Where the aquifer extends considerable distances in all directions from the site of a proposed well field, the most desirable arrangement is one in which the wells are located at equal distances on the circumference of a circle. This pattern equalizes the amount of interference suffered by each well. It should be obvious that a well placed in the center of such a ring of wells would suffer greater interference than any of the others when all are pumped simultaneously. Such centrally placed wells should be avoided in well field layouts.

Where a known source of recharge exists near a proposed site the wells may be located in a semi-circle or along a line roughly parallel to the source. The latter arrangement is the one often used to induce recharge to an aquifer from an adjacent stream with which it is connected. This is a very useful technique in providing an adequate water supply to a small community long after the stream level becomes so low that only an inadequate quantity of poor quality water can be obtained directly from the stream. This is possible since the use of wells permits the withdrawal of water from the permeable river bed and the quality is enhanced by the filtering action of the aquifer materials.

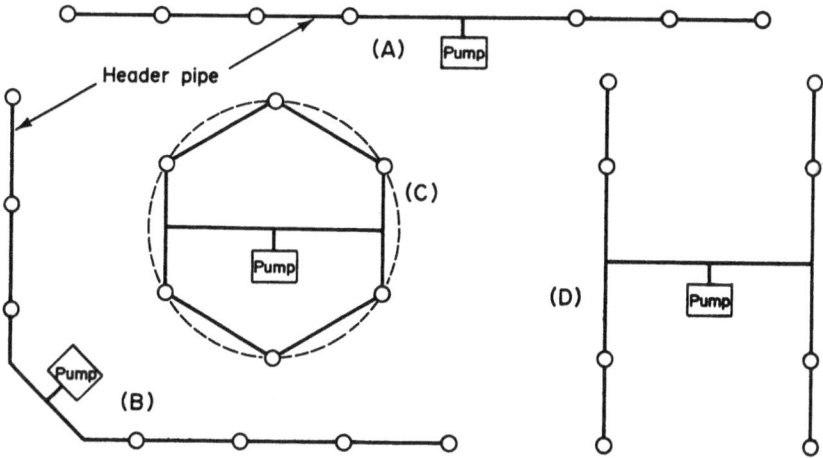

Fig. 2.17 LAYOUT PATTERNS FOR MULTIPLE WELL SYSTEMS USED AS WATER SUPPLY SOURCES. CENTRALLY LOCATED PUMP EQUALIZES SUCTION LIFT.

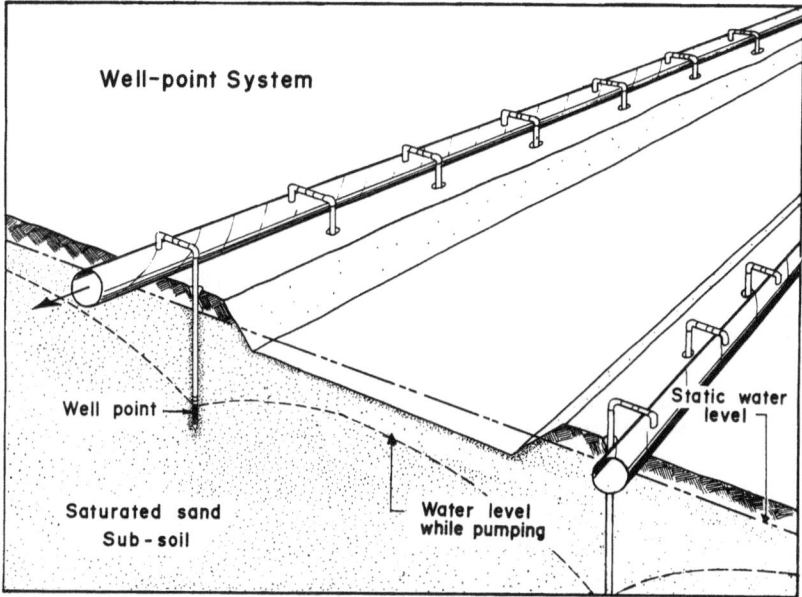

Fig. 2.18 WELL-POINT DEWATERING SYSTEM.

Multiple well or well-point systems are also used on engineering construction sites for de-watering purposes, i.e. to extract water from an area to provide dry working conditions (Fig. 2.18). The significant difference between this use and that for water supplies is the fact that it is now important to create interference in order to lower water levels as much as possible. Closer well spacings than those recommended for water supply purposes are, therefore, necessary. Well spacings for de-watering systems usually range from 2 to 5 feet depending upon the permeability of the saturated sand, the depth to which the water table is to be lowered and the depth to which the well points can be installed in the formation. It is important to note that the de-watering process may require as much as a day of pumping before excavation can begin and must be continued throughout the excavation. Nevertheless, de-watering has often proved more economical than pumping from within a sheet pile surrounded working area.

QUALITY OF GROUND WATER

Generally, the openings through which water flows in the ground are very small. This considerably restricts the rate of flow while at the same time providing a filtering action against particles originally in suspension in the water. These properties, it will be seen, considerably affect the physical, chemical, and microbiological qualities of ground water.

Physical Quality

Physically, ground water is generally clear, colorless, with little or no suspended matter, and has a relatively constant temperature. This is attributable to its history of slow percolation through the ground and the resulting effects earlier mentioned. In direct contrast, surface waters are very often turbid and contain considerable quantities of suspended matter, particularly when these waters are found near populous areas. Surface waters are also subject to wide variations of temperature. From the physical point of view, ground water is, therefore, more readily usable than surface water, seldom requiring treatment before use. The exceptions are those ground waters which are hydraulically connected to nearby surface waters through large openings such as fissures and solution channels and the interstices of some gravels. These openings may permit suspended matter to enter into the aquifer. In such cases, tastes and odors from decaying vegetation may also be noticeable.

Microbiological Quality

Ground waters are generally free from the very minute organisms (microbes) which cause disease and which are normally present in large numbers in surface waters. This is another of the benefits that result from the slow filtering action provided as the water flows through the ground. Also, the lack of oxygen and nutrients in ground water makes it an unfavorable environment for disease-producing organisms to grow and multiply. The exceptions to this rule are again provided by the fissures and solution channels found in some consolidated rocks and in those shallow sand and gravel aquifers where water is extracted in close proximity to pollution sources, such as privies and cesspools. This latter problem has been dealt with in more detail in Chapter 9, where the sanitary protection of ground-water supplies is discussed. Poor well construction can also result in the contamination of ground waters. The reader is referred to the section in Chapter 4 dealing with the sanitary protection of wells.

The solution of the potable water supply problems of Nebraska City, Nebraska, U.S.A. in 1957 bears striking testimony to the benefits derived from percolation of water through the ground and the general advantages of a ground-water supply over one from a surface source. For more than 100 years prior to 1957, Nebraska City depended upon the Missouri River for its domestic water supply. The quality of the water in the river deteriorated as the years went by due to the use of the river for sewage and other forms of waste disposal. To the old problems of high concentrations of suspended matter, dark coloration from decayed vegetation and highly variable temperatures (too warm in summer and too cold in winter) was added bacterial pollution. So bad was this situation that the Missouri River, in this region, soon became recognized as a virtual open sewer and the water no longer met the requirements of the United States Public Health Service Drinking Water Standards for waters suitable to be treated for municipal use.

The search for a new source of supply for Nebraska City led to the use of wells drilled into the sands that underlie the flood plain of the Missouri River at depths up to 100 feet. Wells drilled a mere 75 feet from the river's edge and drawing a considerable percentage of their water from the river yielded a very high quality, clear water that showed no evidence of bacterial pollution or noticeable temperature variation. The lessons of Nebraska City can be put to beneficial use in many other areas of the world.

Chemical Quality

The chemical quality of ground water is also considerably influenced by its relatively slow rate of travel through the ground. Water has always been one of the best solvents known to man. Its relatively slow rate of percolation through the earth provides more than ample time for many of the minerals that make up the earth's crust to be taken into solution. These minerals have varying rates of solution in water, depending upon a number of conditions which themselves may vary widely within a small region. As a result, there may be appreciably wide variations in the chemical quality of ground water found in regions of relatively limited areal extent.

The uses to which ground water can be put depend on its mineral content. Where this content exceeds the recommended limit, treatment should be provided to remove the excessive amounts of the mineral concerned. There are satisfactory methods available for the removal of excessive quantities of the important minerals usually found in ground waters. Expert technical advice should always be sought on the need for and use of these methods.

The mineral content of water is most commonly expressed in parts per million (ppm) which means the number of parts, by weight, of the mineral found in one million parts of the solution. For example, a concentration of

10 ppm of iron means that in every million pounds (or kilograms) of the water examined there will be found 10 pounds (or kilograms) of iron. Another very common form of expression is that of milligrams per liter (mg/l or mg per l) which is the number of milligrams of the mineral found in one liter of water. This latter unit differs so little from the former that they are, for all practical purposes, considered equal and are commonly used interchangeably.

The following are among the more important chemical substances and properties of ground waters which are of interest to the owners of small wells: iron, manganese, chloride, fluoride, nitrate, sulfate, hardness, total dissolved solids, pH, and dissolved gases such as oxygen, hydrogen sulfide, and carbon dioxide.

Iron and *manganese* are usually considered together because of their resemblances in chemical behavior and occurrence in ground water. It is important to note that iron and manganese, in the quantities usually found in ground water, are objectionable because of their nuisance values rather than as a threat to man's health. They both cause staining (reddish brown in the case of iron and black in the case of manganese) of plumbing fixtures and clothes during laundering. Iron deposits may accumulate in well screens and pipes, restricting the flow of water through them. Iron-containing waters also have a characteristic taste which some people find unpleasant. Such waters, when first drawn from a tap or pump, may be clear and colorless, but upon allowing the water to stand, the iron settles out of solution giving a cloudy appearance to the water and later accumulating in the bottom as a rust-colored deposit.

Chlorides occur in very high concentrations in sea water, usually of the order of 20,000 mg/l. Rainwater, however, contains much less than 1 mg/l of chloride. Aquifers containing large chloride concentrations are usually coastal ones directly connected to the sea or which were so connected some time in the past. Excessive pumping of wells in aquifers directly connected to the sea or to brackish-water rivers will cause these high chloride-containing waters to move into the otherwise fresh water zones of the aquifers. Expert technical advice should be sought on the possibility of such an occurrence.

Water with a high chloride content usually has an unpleasant taste and may be objectionable for some agricultural purposes. The level at which the taste is noticeable varies from person to person but is generally of the order of 250 mg/l. A great deal depends, however, on the extent to which people have been accustomed to using such waters. Animals usually can drink water with much more chloride than humans can tolerate. Cattle have, reportedly, been known to consume water with a chloride content ranging from 3000 mg/l to 4000 mg/l.

Fluoride concentrations in ground water are usually small and mainly derived from the leaching of igneous rocks. Notable among the few cases of high concentrations is the reported 32 mg/l from a flowing well near San Simon, Arizona, U.S.A. High concentrations have also been reported in some parts of India, Pakistan and Africa.

When present in concentrations less than 1.0 mg/l in water, fluoride generally reduces tooth decay in small children and is desirable. Excessive concentrations, however, result in a brown discoloration and pitting of the teeth called dental fluorosis. This condition is particularly noticeable in children but can also occur in adults. The level of concentration at which this adverse effect occurs varies from one community to another depending upon factors such as temperature and fluoride intake to the body through food. It is also likely that continued consumption of waters containing fluoride in excess of 4 mg/l may affect bone structure. Waters with concentrations in excess of about 3.5 mg/l are usually not recommended for drinking water supplies.

Nitrate content in ground waters varies considerably and is often unrelated to the rock formations in the area. High nitrate concentrations are very often due to the percolation of surface waters containing human wastes and/or animal and other agricultural waste products into aquifers or to the direct flow of contaminated surface runoff into wells. Precautions must therefore be taken in the location and construction of shallow wells in areas where privies, cesspools and barnyards are to be found. These precautions are discussed in later sections on well design (Chapter 4) and the sanitary protection of ground-water supplies (Chapter 9).

High concentrations of nitrate in water produce an effect known as cyanosis (methemoglobinemia) in infants. This condition which is characterized by a bluish discoloration of the skin, listlessness and drowsiness can be fatal. For this reason, water containing nitrate in excess of 45 mg/l should not be used in preparing food for babies under six months of age. It should be noted that the boiling of such water will only serve to increase the nitrate concentration.

Sulfate in ground water is derived mainly from the leaching of natural deposits of magnesium sulfate (Epsom salts) or sodium sulfate (Glauber's salt) both of which, in sufficient quantities, may produce laxative effects.

Hardness is that property of water best demonstrated by the readiness with which it dissolves soap to produce suds. No suds are produced in a hard water until the minerals causing the hardness have been removed by chemical combination with constituents of the soap. The greater the hardness, the more soap is required to produce suds.

The hardness produced by the bicarbonates of calcium and magnesium can be virtually removed by boiling the water and is called *temporary hardness.* The hardness caused mainly by the sulfates and chlorides of calcium and magnesium cannot be removed by boiling and is called *permanent hardness.* *Total hardness* is the sum of the temporary and permanent hardness.

The removal of temporary hardness by heat causes the deposition of calcium and magnesium carbonates as a hard scale in kettles, cooking utensils, heating coils, and boiler tubes which may result in a waste of fuel.

Total dissolved solids refer to the sum total of all the minerals such as chlorides, sulfates, etc. found dissolved in the water. A water with a high total dissolved solids content would therefore be expected to present the taste, laxative and other problems associated with the individual minerals. Such waters are usually corrosive to well screens and other parts of the well structure.

pH is a measure of the hydrogen ion concentration in water and indicates whether the water is acid or alkaline. It ranges in value from 0 to 14 with a value of 7 indicating a neutral water, values between 7 and 0 increasingly acid and between 7 and 14 increasingly alkaline waters. Most ground waters in the United States have pH values ranging from about 5.5 to 8. Determination of the pH value is important in the control of corrosion and many processes in water treatment.

The *dissolved oxygen* content of ground waters is usually low particularly in waters found at great depths. Oxygen speeds up the corrosive attack of water upon iron, steel, galvanized iron, and brass. The corrosive process is also more rapid when the pH is low.

Hydrogen sulfide is recognizable by its characteristic odor of rotten eggs. It is very often found in ground waters which also contain iron. In addition to the odor, which is noticeable at as low a concentration as 0.5 mg/l, hydrogen sulfide combines with oxygen to produce a corrosive condition in wells and also combines with iron to form a scale deposit of iron sulfide in pipes. Most of the hydrogen sulfide can be removed from ground water by spraying it into the air or allowing it to cascade in thin layers over a series of trays.

Carbon dioxide enters water in appreciable quantities as the water percolates through soil in which plants are growing. Dissolved in water, it forms carbonic acid which, together with the carbonates and bicarbonates, controls the pH value of most ground waters. A reduction of pressure, such as caused by the pumping of a well, results in the escape of carbon dioxide and an increase in the pH value of the water. Testing of ground-water samples for carbon dioxide content and pH, therefore, requires the use of special techniques and should be done at the well site. The escape of carbon dioxide from a water may also be accompanied by the settling out of calcium carbonate deposits.

While the above list includes those chemical substances that are likely to be of greatest general concern to owners of small wells, it is by no means an exhaustive one nor intended to be such. Conditions peculiar to specific areas may require analyses of ground waters for other substances. The group of elements often referred to as the *trace elements* because of the very low concentrations in which they are usually found in water are here worth mentioning. Among these are arsenic, barium, cadmium, chromium, lead and selenium, all of which are considered toxic to man at very low levels of intake (the order of a fraction of 1 mg/l). Since the rate of passage of some of these elements through the body is very slow, the effects of repeated doses are additive and chronic poisoning occurs.

Trace elements generally are not present in objectionable concentrations in ground waters but may be so in a few specific areas. It has been reported for example, that arsenic has been found in sufficiently high concentrations in ground waters in some parts of Argentina and Mexico to be considered injurious to health. Problems are most likely to arise in areas where waste discharges from industries, such as electro-plating, and overland run-off containing high concentrations of pesticides (insecticides and herbicides) enter aquifers.

The presence of these trace elements in drinking water are generally not detectable by taste or smell or physical appearance of the water. Proper chemical analyses are required for their detection. Health departments, laboratories, geological survey departments, and other competent agencies should be consulted in areas where waste disposal is likely to increase the natural content of these elements in ground water or where the natural levels are likely to be high because of the local geology.

CHAPTER 3

GROUND-WATER EXPLORATION

Water can be found almost anywhere under the earth's surface. There is, however, much more to ground-water exploration than the mere location of subsurface water. The water must be in large quantities, capable of sustained flow to wells over long periods at reasonable rates, and of good quality. To be reliable, ground-water exploration must combine scientific knowledge with experience and common sense. It cannot be achieved by the mere waving of a magic forked stick as may be claimed by those who practice what is variously referred to as water witching, water dowsing, or water divining.

Finding the right location for a well that produces a good, steady water supply all the year round is usually the job of scientists trained in hydrology. These scientists are called hydrologists. Their help may be sought from geological survey departments, governmental and private engineering organizations, and universities if and when available. These experts should always be consulted for large scale ground-water development schemes because of the great capital expenditures usually involved. However, it should be apparent from the remaining sections of this chapter that a sufficient number of the tools of the hydrologist is based upon the application of common sense, intelligence and good judgement to permit their reasonably successful use by the average individual interested in the location of small wells. The interpretation of geologic data may present problems though, with some help, these need not be totally insurmountable to some of our readers. The use of well inventories and surface evidence of ground water location should be much less difficult and find greater general application.

The following sections describe the simpler tools of the hydrologist and his use of them. The more sophisticated methods of exploration involving the use of geophysics are considered beyond the scope of this manual and, therefore, have been excluded. It is sufficient to note that they are available to the hydrologist to provide him with additional information on which to base his selection.

GEOLOGIC DATA

Before visiting the area to be investigated, the hydrologist seeks out and studies all available geologic data relating to it. These would include geologic maps, cross-sections and aerial photographs.

Geologic Maps

Geologic maps, of which Fig. 3.1 is an example, show where the different rock formations, consolidated or unconsolidated, come to the land surface or outcrop, their strike or the direction in which they lie, and their dip or the angle at which they are inclined to the horizontal. Other useful information

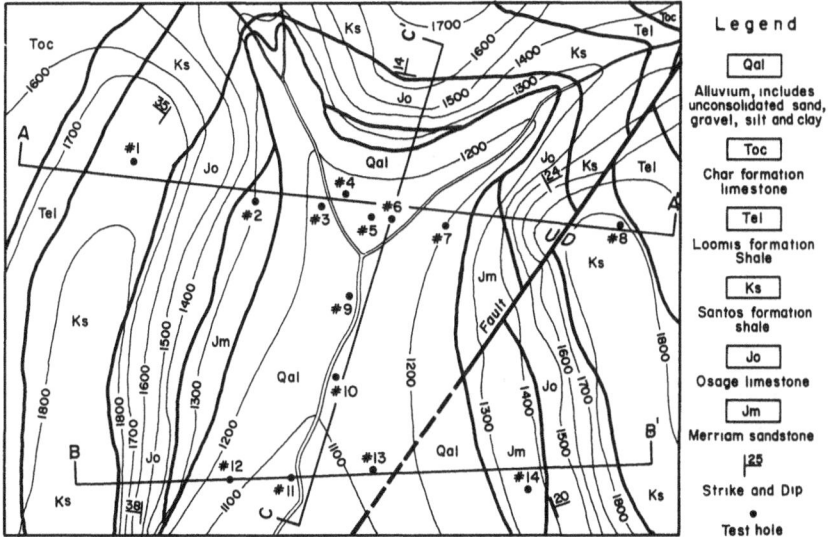

Fig. 3.1   EXAMPLE OF A GEOLOGIC MAP SHOWING TEST HOLE LOCATIONS.

shown would include the location of faults and contour lines indicating depth to bedrock throughout the area. Faults are lines of fracture about which the rock formations are relatively dislocated. They are the result of forces acting in the earth to cause lateral thrust, slippage or uplift. The hydrologist can determine the location and areal extent of aquifers from the type and location of rock outcrops and the location of faults. Faults are also likely sites for the occurrence of springs. The width of the outcrop and angle of dip indicate to him the approximate thickness of an aquifer and the depths to which it can be found. The combination of strike and dip tell him in which direction he should locate a well to obtain the maximum thickness of the aquifer. The surface outcrops also indicate the possible areas of recharge to an aquifer and, by deduction, the direction of flow in the aquifer. The bedrock contours indicate the maximum depth to which a well should be drilled in search of water.

Geologic Cross-Sections

Geologic cross-sections provide some of the main clues to the ground-water conditions of a locality. They indicate the character, thickness, and succession of underlying formations and, therefore, the depths and thicknesses of existing aquifers. The main sources of information for the preparation of these sections are well records and natural exposures where the rock faces have not been greatly altered by weathering. Examples of the latter

29

may be seen in some river valleys and gorges. These sections may also indicate whether water-table or artesian conditions exist in an aquifer. The cross-sections of Fig. 3.2, drawn from the geologic map of Fig. 3.1, illustrate many of the important features mentioned above.

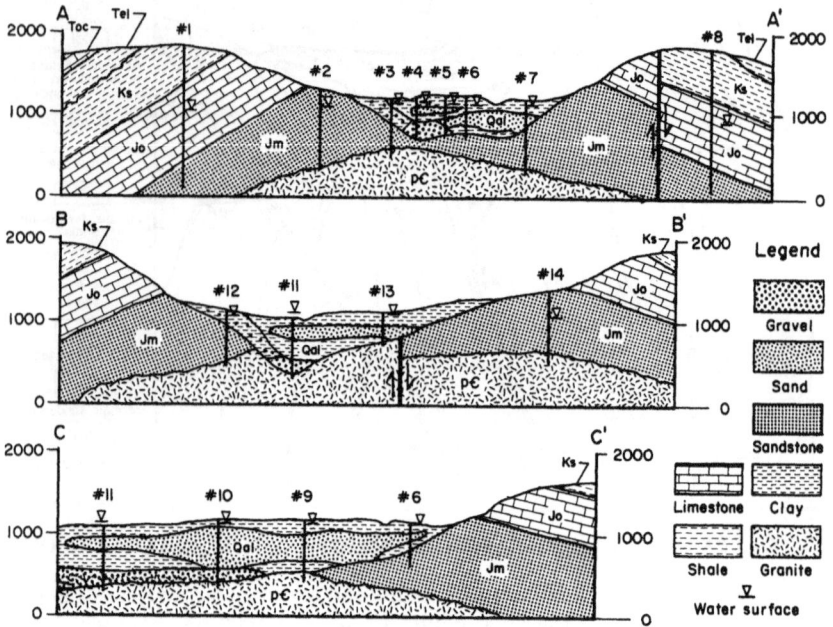

Fig. 3.2 GEOLOGIC CROSS-SECTIONS FROM THE MAP OF FIG. 3.1.

Aerial Photographs

Aerial photographs, skillfully interpreted, provide valuable information on terrain characteristics which have considerable bearing on the occurrences of ground water. Features which indicate subsurface conditions such as vegetation, land form and use, erosion, drainage patterns, terraces, alluvial plains, and gravel pits are apparent on aerial photographs. The skillful interpreter of aerial photographs can determine the most promising areas for ground-water development.

INVENTORY OF EXISTING WELLS

The hydrologist next makes a study of all available information on existing wells.

Well logs which are records of information pertaining to the drilling and construction of wells would be the main sources of information. From these logs he may obtain such information as the location and depth of the well;

depth, thickness, and description of rock formations penetrated; water level variations as successive strata are penetrated; yields from water-bearing formations penetrated and the corresponding drawdowns; the form of well construction; and the yield and drawdown of the well upon completion. Many drilling organizations also keep samples of rocks from the various formations penetrated. Associated with the well log should be a record of the water quality (physical and chemical characteristics) of water-bearing strata encountered. Of further interest to the hydrologist would be records of any tests making use of the well or materials from the well to determine the hydraulic characteristics such as permeability and transmissibility of the aquifer. To complete the picture, he would be interested in records of the variations in yield and water quality and a history of any problems associated with the well since its completion. The hydrologist may wish to have new checks made on some aspects of these records such as the water quality and yield.

All these records may not be available from any single source. In addition to the various agencies already mentioned, the hydrologist may have to consult well owners and drilling organizations.

With records from a sufficient number of wells, the hydrologist would now be in a position to make a contour map of the water-table or upper surface of the zone of saturation. To do this, he uses the measured depths from land surface to the water table at wells and the height of the land surface relative to sea level which he obtains from topographic maps or a site survey. He then connects all points of equal elevation of the water table on a map. This contour map shows the shape of the water surface. It is a very important map in that it shows not only the depth below which ground water is stored but also, from the slope of the water table, the direction in which the water moves.

SURFACE EVIDENCE

The hydrologist is now ready to visit the area and take a closer look at any surface evidence of ground-water occurrence. He will examine in greater detail the important superficial features he had noted on the topographic maps and aerial photographs. Among the features that would provide valuable clues would be land forms, stream patterns, springs, lakes, and vegetation.

Ground water is likely to occur in larger quantities under valleys than under hills. Valley fills containing rock waste washed down from mountain sides are often found to be very productive aquifers. The material may have been deposited by streams or sheet floods with some of the finer material getting into lakes to form stratified lake beds. Some of these deposits may also be found to have been transported by wind and re-deposited as sand dunes. All these and other factors influence the rate at which the valley fill will yield water. Coastal terraces, formed by the sinking and raising of coastal areas relative to sea level in the geologic past, and coastal and river plains are other land forms that would indicate the presence of good aquifers.

Any evidence of surface water such as streams, springs, seeps, swamps, or lakes is a good indication of the presence of some ground water, though not

31

necessarily in usable quantity. The sand and gravel deposits found in river beds may very often extend laterally into the river banks which may be penetrated by shallow, highly productive wells.

Vegetation, particularly water-loving types when found in arid regions, provides good clues to shallow ground-water occurrence. Evidence of unusually thick overgrowth is generally a sure pointer to the presence of streams and other surface waters, the vicinities of which would be likely sites for ground-water investigations. Fig. 3.3 demonstrates the application of some of the principles outlined above in the selection of possible well sites.

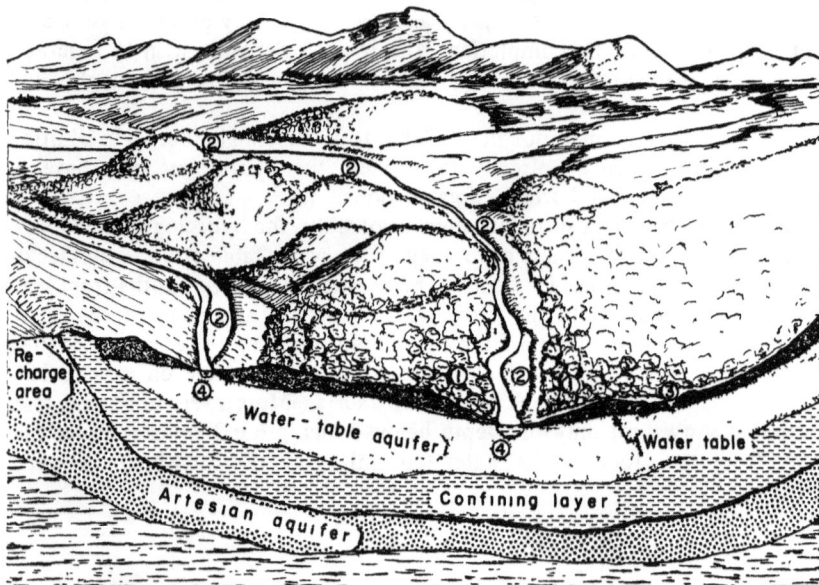

Fig. 3.3 SURFACE EVIDENCE OF GROUND-WATER OCCURRENCE. (Adapted from Fig 4, *Water Supply For Rural Areas And Small Communities*, WHO Monograph Series No. 42, 1959.)
   1 – Dense vegetation indicating possible shallow water table and proximity to surface stream.
   2 – River plains: possible sites for wells in water-table aquifer.
   3 – Flowing spring where ground water outcrops. Springs may also be found at the foot of hills and river banks.
   4 – River beds cut into water-bearing sand formation. Indicate possibility of river banks as good well sites.

In many areas, some of the records, maps, and other relevant information so far discussed will not be available to the hydrologist. Where the size of the project warrants it, he will arrange to fill in the information as far as is practicable. In other situations, very likely the case for the size of wells herein considered, he will simply use his best judgement based on the readily available information.

32

CHAPTER 4

WATER WELL DESIGN

Generally, the aim of engineering design is to achieve the best possible combination of performance, useful life and reasonable cost. The designer of small wells will often find that his optimum solutions involve a variety of compromises and that he must adopt a flexible approach to each problem. Among these compromises is the need to sacrifice performance or efficiency in order to reduce costs. For example, in the situation where a small yield is required from a very thick and permeable aquifer, a less efficient type of intake section such as slotted pipe may justifiably be used in a small well to save the extra cost of a more efficient factory-manufactured screen. Here, the limited yield relative to the highly productive nature of the aquifer makes cost and availability of funds assume a more important role than hydraulic efficiency. It may also be considered worthwhile to compromise the useful life of a small well with respect to its cost. With stainless steel and other non-corrosive materials costing two to three times as much as ordinary steel, a designer may use well casing of the latter material under corrosive conditions, fully expecting to replace it, perhaps in one-half the time he would have had he used stainless steel. He may very well have based his decision on the fact that at the end of the shorter useful life, extra funds might be available for a replacement of the existing well.

For design purposes, a well to be constructed in unconsolidated materials may be considered as consisting of two main parts. The upper part or *cased section* of the well serves as housing for the pumping equipment and as a vertical conduit through which water flows from the aquifer to the pump or to the discharge pipe of a flowing artesian well. It is usually of water-tight construction and extends downward from the surface to the impervious formation immediately above an artesian aquifer or to a safe depth below the anticipated pumping water level (see the later section of this chapter dealing with sanitary protection of wells). It is also referred to as the *well casing*.

The lower or *intake section* of the well is that part of the well structure where water from the aquifer enters the well. The intake section may be simply the open lower end of the well casing, though this would be a most unsatisfactory arrangement in unconsolidated formations. The disadvantages are the large well diameters required for the natural seepage of water into the well and the tendency for aquifer material to heave into the well casing as the well is being pumped. A screening device known as a *well screen* should be used instead. Such a screen permits the use of techniques aimed at increasing the natural seepage rate into the well (see later section on well development), thus making a much smaller well practicable. In addition to ensuring the relatively free entry of water into the well at low velocity, the screen must provide structural support against the collapse of the unconsolidated formation material and prevent the entry of this material with water into the well.

CASED SECTION

The selection of the well casing *diameter* is usually controlled by the type and size of the pump that is expected to be required for the desired or potential yield of the well. The well casing must be large enough to accommodate the pump with sufficient clearance for easy installation and efficient operation. For larger wells, such as those used for municipal and industrial supplies, the casing diameter should be chosen as two nominal sizes (never less than one nominal size) larger than that of the pump bowls. For wells of 4 inches and less in diameter it is satisfactory to select a casing diameter which is one nominal size larger than that of the pump bowls, pump cylinder or pump body. The above assumes the use of a deep-well type of pump which is usually suspended by pipe column and/or shaft within the well casing. A pump having a bowl diameter (see Fig. 8.11) greater than 3 inches should not, according to this rule, be installed in a 4-inch diameter casing.

In small wells where pumping water levels below ground surface are known to be within the practical suction limits (15 feet or less) of most surface-type pumps, such pumps are either directly connected to the top of the well casing or connected to a suction pipe suspended inside the well casing. The well casing diameter may then be selected in relation to the diameter of the suction or inlet of the pump, bearing in mind that it is not good practice to restrict the suction capacity of the pump by using pipe of a smaller diameter than that of the suction side of the pump.

In larger and deeper wells than those being considered, it is sometimes advantageous for economic and other reasons to reduce the casing diameter at levels below the lowest anticipated pumping depth. This is done by telescoping one or more smaller sized casing sections through the uppermost one. This saves the extra cost of extending the large diameter casing all the way down to the aquifer when a smaller size of pipe would be sufficient to accommodate the anticipated flow with reasonable head loss. However, there is little justification for this type of design in wells of 4 inches and less in diameter and not more than 100 feet deep.

INTAKE SECTION
Type and Construction of Screen

The single factor with greatest influence on the efficient performance of a well is the design and construction of the well screen. A properly designed screen combines a high percentage of open area for relatively unobstructed flow into the well with sufficient strength to resist the forces to which the screen may be subjected both during and after installation in the well. The screen openings should preferably be shaped so as to facilitate flow into the well while making it difficult for small particles to become permanently lodged in them and thus restrict flow. A discussion of various types of well screens and their uses is presented in the following paragraphs.

The *continuous-slot* type of well screen shown in Fig. 4.1 is made with cold-drawn wire, approximately triangular in section, wound spirally around a circular array of longitudinal rods. The wire is welded to the rods at all points

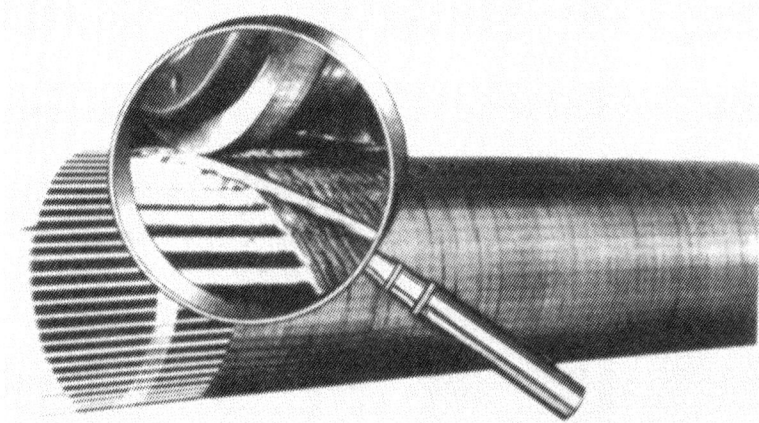

Fig. 4.1 FABRICATION OF A CONTINUOUS-SLOT TYPE OF WELL SCREEN.

at which they cross. The resulting cylindrical well screen becomes a one-peice, rigid unit.

The stronger the material used in construction, the smaller would be the dimensions of the wire rods and hence the greater the ratio of open area to solid area of the screen surface. These screens are being made of metals such as galvanized iron, steel, stainless steel and various types of brass. Experiments are also in progress with the use of plastic materials.

The percentage of open area is the factor exerting the greatest influence on the efficiency of a screen. As will be shown later, the size of the well screen opening is determined from the size of the particles of the material composing the aquifer. With this size fixed, the aim of screen design is to obtain the maximum possible total open area in a given length of screen. The greater the total open area, the lesser is the resistance to flow into the well. The entrance velocity through the larger intake area is also lower and so is the resulting head loss for flow through the screen. Hence we have a more efficient well screen. The greater the percentage of open area in a screen, the greater is the total open area in a given length of screen.

Looking at it in another way, the greater the percentage open area of a screen, the shorter is the length of screen required for a given rate of flow at a given velocity. This means that a saving in construction costs can be made through the use of a shorter length of screen. The continuous-slot type of screen provides more intake area per square foot of screen surface or per unit length of screen than any other known type and, therefore, can result in savings when used.

Along with maximum open area in a well screen, the design must also be such that the openings do not become clogged by sand particles after the screen is placed in the aquifer. This is achieved by the use of V-shaped openings formed by the triangular shaped wire as shown in Fig. 4.2. In Fig. 4.3 is shown a sand grain entering and passing through a V-shaped opening, never clogging it, while remaining in other known types of openings to clog them. This property of the V-shaped opening is of special importance when developing the well, as the developing process is based on passing the smaller sizes of sand particles through the screen and removing them from the well. This process, a necessary one for the completion of the well, is described later in this chapter.

Fig. 4.2 **SECTION OF CONTINUOUS-SLOT TYPE SCREEN SHOWING V-SHAPED OPENINGS.**

Another notable feature of the continuous-slot type of screen is the fact that the slot openings can be easily varied in size even within the same section of screen if the geologic conditions so require. This is done simply by altering the set spacing at which the adjacent wires are wrapped. Thus a single section of screen can be made with one or more different sizes of slot openings. The width of slot openings can also be held to close tolerances.

Continuous-slot well screens are made with practically any width of opening 0.006 inch and larger. The slot openings are designated by numbers corresponding to the width of the opening in thousandths of an inch. Thus a screen with a No. 10 slot has openings 0.010 inch wide.

Louver- or *shutter*-type well screens have rows of openings in the form of shutters (Fig. 4.4). Manufacturers can and do arrange the openings either at right angles or parallel to the axis of the screen. The openings are produced in the wall of a welded tube by a stamping operation using a die. The range of sizes of openings is limited by the sizes of the set of dies used by each manufacturer. An unlimited range of die sizes would not be practical. This is one deficiency of this type of screen by comparison with the continuous-slot. Another important deficiency is the much lower percentage of open area in shutter-type screens. This is so because sizeable blank spaces must be left between adjacent openings if the metal is not to be torn in the stamping process.

Yet another shortcoming of the shutter-type screen is the tendency of the openings to become blocked during the development of wells (Fig. 4.3) where the aquifer material contains an appreciable proportion of sand. This type of screen is, therefore, best used in artifically gravel-packed wells, a description of which is presented later ın this chapter.

36

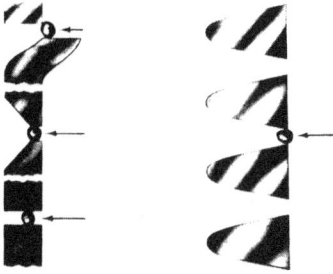

Fig. 4.3 THE V-SHAPED OPENINGS OF THE CONTINUOUS-SLOT TYPE OF SCREEN (RIGHT) ALLOWS S A N D GRAINS BARELY SMALLER THAN THE WIDTH OF THE OPENINGS TO PASS FREELY WITHOUT CLOGGING. OPENINGS WITHOUT THE TAPER TEND TO HOLD PARTICLES JUST SMALL ENOUGH TO ENTER THEM.

The *pipe-base* well screen is another type of screen in use. It consists of a jacket around a perforated metal pipe. The jacket may be in the form of a trapezoidal-shaped wire wound directly onto and around the pipe (called a wrapped-on-pipe screen). Alternatively the wire may be wound over a series of longitudinal rods spaced at fixed intervals around the circumference of the pipe. The latter is a more efficient type of screen as the rods hold the wire away from the pipe surface to reduce the blocking of the screen openings. A stronger screen can be obtained by using a slip-on jacket made of an integral unit of welded well screen.

The perforations or holes in the pipe and the spaces between adjacent turns of the wrapping wire form two sets of openings in this type of screen. Usually the total open area of the holes in the pipe is less than that between the wrapping wire. It is, therefore, the holes in the pipe that control the performance of the screen. The percentage open area in the pipe is usually low and hence this type of screen is relatively inefficient.

Very often this type of construction is used in order to avoid making a screen entirely of the costly non-corrosive alloys such as stainless steel, bronze or brass. Such alloys are then used only in the jacket while the pipe is of steel. A screen so constructed with two or more metals would be subject to failure from galvanic corrosion. Construction of the screen entirely of one of the non-corrosive alloys, while being more costly, will solve this problem and result in a more durable screen.

Drive points or *well points*, as they are commonly known, are short lengths of well screen which are attached to successive lengths of pipe and driven by repeated blows to the desired position in an aquifer or in a formation to be dewatered. A forged steel point is usually attached to the lower end to facilitate penetration into the ground.

Well points are made in a variety of types and sizes. Most commonly, they are designed for direct attachment to either 1¼-inch or 2-inch pipe. They can be made of the *continuous-slot* type of well screen (Fig. 4.5), thus benefitting from all the desirable features of that type of screen. Such screens will withstand hard driving, but care should be taken to avoid twisting them while driving.

A common type of well point is the *brass jacket* type. It consists of a perforated pipe covered with bronze wire mesh which is, in turn, covered with a perforated brass sheet to protect it from damage. The pointed lower

37

Fig. 4.4 LOUVER- OR SHUTTER-TYPE WELL SCREEN, BEST USED IN ARTIFICIALLY GRAVEL-PACKED WELLS. (From Layne and Bowler, Inc., Memphis, Tennessee.)

end, made of forged steel, carries a wider shoulder to protect the screen from damage by gravel or stones while being driven. The limitations of pipe-base screens also apply to this type of well point.

Another type of well-point construction is the *brass tube* type consisting of a slotted brass tube slipped over perforated pipe. It has an advantage over the wire-mesh jacket type in that it is not as easily ripped or damaged.

The sizes of openings for the continuous-slot type of well points are designated as described for the continuous-slot well screens. Mesh-covered well point openings are designated by the mesh size in terms of the number of openings per linear inch. The common sizes are 40, 50, 60, 70 and 80 mesh.

Slotted pipe is sometimes used as a substitute for well screens particularly in the smaller sized wells under consideration in this manual. The openings or slots in the pipe are usually cut with a sharp saw, electrically operated if possible, to maintain accuracy and regularity in size. Several other methods have been used, however, such as cutting with an oxyacetylene torch and punching with a chisel and die or casing perforator.

The method of construction immediately suggests a number of important limitations to the use of slotted pipe as well screens. These are: (1) structural strength requires wide spacing of slots, resulting in a low percentage of open area; (2) openings may be inaccurate, varying in size throughout the length of each slot; (3) openings narrow enough to control fine sands are difficult, if not impossible, to produce; (4) the lack of continuity of the openings reduces the efficiency of the process of well development; and (5) the slotting and perforation of steel pipe makes it more readily subject to corrosion, particularly at the jagged edges and surfaces.

Slotted plastic pipe has been finding increasing use in small diameter wells in recent years. Its light weight and ease of handling make it suitable for use in remote areas not easily reached by motor driven vehicles. It is non-corrosive and less costly than steel pipe in sizes 4 inches in diameter and

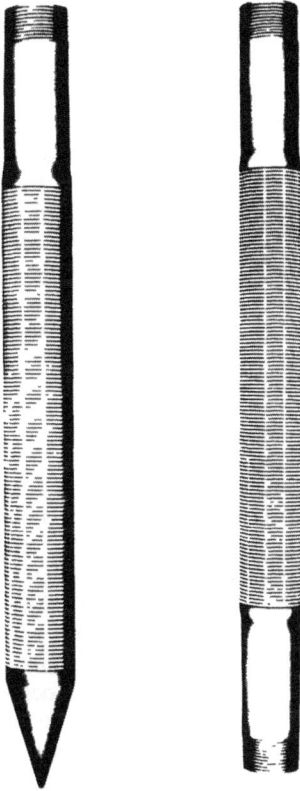

Fig. 4.5 CONTINUOUS-SLOT TYPE
OF WELL POINT AND EX-
TENSION SECTION.

smaller. In addition, the slots can be easily made on location with a sharp saw within reasonable limits of accuracy. Slots cut spirally around the circumference of the pipe in the manner shown in Fig. 4.6 will result in less weakening of the pipe and closer spacing of the slots than if they were made at right angles to the axis. Consequently, the percentage of open area is greater. Slots made at right angles to the axis of plastic pipe are subject to tearing at both ends if the slotted pipe is bent when handling it during installation. This tendency is reduced by the use of the spiral design.

The most convenient type of joint for use with small diameter plastic pipe in well construction is the spigotted joint. For these joints, the manufacturers supply a quick-setting cement which provides more than adequate and lasting strength. The slotted plastic-pipe screen can be lowered into a previously drilled hole on the end of casing of the same material. Steel clamps are used to suspend the string of pipes while adding new lengths. It may also be washed, open ended, with a jet of water into a previously drilled hole. Suitable drilling mud should be used during rotary drilling operations to prevent the open hole from collapsing while the string of plastic pipe is being placed in position. Care should be taken to wash the hole clear of all cuttings before placing the pipe. Plastic pipe generally requires the use of greater care during handling and installation operations than do metal pipes.

It cannot be contended that slotted plastic pipe will be as efficient a well screen as the continuous-slot type. However, when only small quantities of water are required from relatively thick (20 feet and greater) sand and gravel or gravel aquifers, efficiency loses some of its importance to economy and ease of construction. Under these conditions, together with the ones already mentioned, slotted plastic pipe is an attractive alternative to the continuous-slot or other manufactured type of well screen. It is particularly suited to the provision of individual water supplies in remote and inaccessible areas.

Screen Length, Size of Openings and Diameter

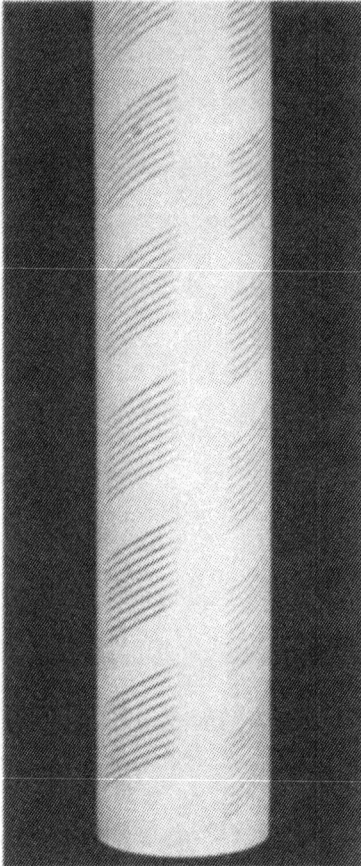

Fig. 4.6   SLOTTED PLASTIC PIPE.

The length, size of openings and diameter of the well screen are the remaining design features which influence the efficiency of flow into a well. Together, they determine the entrance velocity of flow through the screen into the well. This entrance velocity in turn influences the head or pressure loss required for maintaining the flow and, as a consequence, also influences the efficiency of the screen for that rate of flow.

If designing a well to obtain the maximum yield from an aquifer, then the procedure would first be to select the screen length and size of openings based on the natural characteristics of the aquifer. The screen diameter would then be selected so as to provide enough total area of screen openings that the entrance velocity does not exceed the chosen design standard. Usually, however, small wells are designed to provide a certain limited yield, well below the maximum possible yield, and the screen diameter is first chosen essentially with a view to keeping costs down to a minimum. The diameter selected would then be the smallest practicable one, consistent with the expected yield and the diameter of the casing. Normally, it is not considered good practice to use a well screen of larger diameter than that of the casing. The size of the screen openings is, as before, fixed by the aquifer characteristics, but the screen length is, in this case, determined by the total area of screen openings required to keep the entrance velocity at or below the design standard. Should the screen length determined on this basis be greater than the thickness and other characteristics of the aquifer would permit, then the screen length is chosen as the maximum consistent with these limitations. Following this, a suitable diameter is chosen to be consistent with the design standard for entrance velocity into the screen. A more detailed discussion of the design standard for the entrance velocity follows discussions of the effects of aquifer characteristics on the selection of screen length and size of openings.

Manufacturers make well screens in two series of sizes, the telescope-size and the pipe-size or ID-size. *Telescope-size* screens are designed to be "telescoped" or lowered through the well casing to the final position. The diameter of each screen is just sufficiently smaller than the inside diameter of the corresponding size of standard pipe to permit the screen to be freely lowered through the pipe.

The *pipe-size* or *ID-size* series of well screens have the same inside diameter as the corresponding size of standard pipe. This type of screen is used when it is desired to maintain the same diameter throughout the full depth of the well. They are provided, in the small sizes under consideration, with either welded or threaded end connections.

Screen length: The screen length selection can be influenced by the thickness of the aquifer. While definite rules may be set, based on this relationship, for large wells it would be unwise to do so for small ones. A farmer or home-owner should not be burdened with a long and costly well screen in a thick aquifer when his requirements are so small as not to warrant it. The screen length should be sufficient to meet his needs with a reasonable drawdown in the well. As already stated, a compromise must be made between well cost and well efficiency. The other extreme must also be avoided. Economization should not be taken to the point where the length of screen provided is such that the yield barely meets the owner's present needs. A reasonable allowance should be made for his future needs. Failure to do so may, in the long run, prove to be far more costly to the owner.

It is important to note that in a thick aquifer, well yield is much more effectively increased by increasing the screen length than by proportionately increasing the screen diameter. Doubling the screen diameter, for instance, will only result in an increase of 10 to 15 percent in the yield. In most cases, however, doubling the screen length will result in the yield being almost doubled. It is, therefore, much better to use screen length as a controlling factor on well yield rather than screen diameter in thick aquifers.

The role played by aquifer characteristics in screen length selection is best demonstrated with the use of a few examples. Where a thick layer of coarse sand or gravel underlies a layer of fine sand as shown in Fig. 4.7A, the screen length should be at least one-third the thickness of the coarse sand layer. For the sitatuions shown in Fig. 4.7B and Fig. 4.7C, almost the entire thickness of the lower layer of coarse sand should be screened. Should this prove inadequate for the desired yield, then it would be necessary to extend the screen a short distance into the overlying finer sand. Where a coarse sand overlies a fine sand as in Fig. 4.7D, it should normally be sufficient to place the screen in the coarse sand layer with the length being equal to about one-half the thickness of that layer.

In thin aquifers confined by clays, particularly clays that tend to be easily eroded when exposed to water, screen lengths should be chosen so as to avoid the possibility of placing screen openings opposite these clays. Screening of clay layers could result in their collapse during the well development process with the well forever producing a muddy water.

41

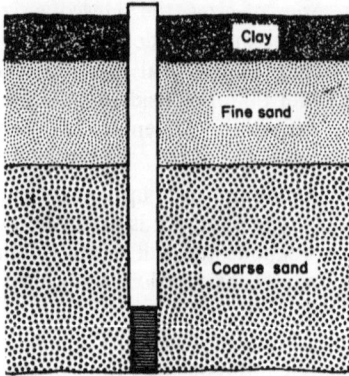

(A) Coarse part of formation is thick

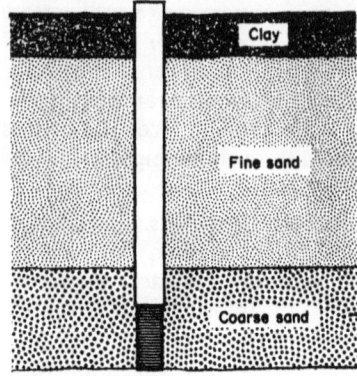

(B) Coarse part of formation is thin

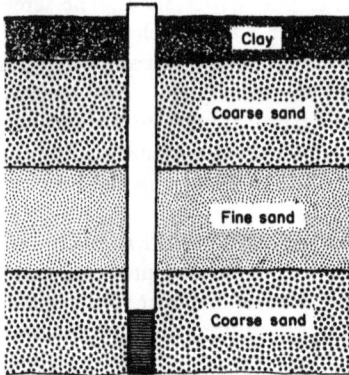

(C) Alternate layers of coarse and fine sand

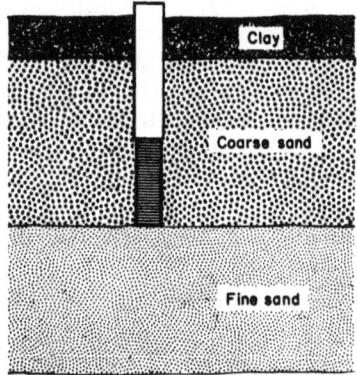

(D) Coarse material above fine sand

Fig. 4.7 RECOMMENDED POSITIONING OF WELL SCREENS IN VARIOUS STRA-
TIFIED, WATER-BEARING SAND FORMATIONS.

Screen slot opening: An understanding of the method of selecting the size of
screen slot openings first of all requires an understanding of the process and
objectives of well development. As previously stated, fine material occupies
part of the otherwise larger pore spaces of water-bearing formations, thus increasing the head losses due to friction and reducing the quantity of water
yielded per unit of drawdown in a well (specific capacity). The object of *well
development* is to remove as much of this finer material as possible from a zone
around the well to improve the specific capacity and efficiency of the well.
There are a variety of methods that are used for inducing the flow of this fine
material through the well screen and then extracting it by pumping or bailing.
Some of these methods are described in Chapter 6. It is sufficient to note at
this point that well development involves the removal of the finer aquifer
material in the vicinity of a well and that this removal takes place through the
screen and out of the casing.

42

The limiting size of material to be removed, therefore, fixes the size of the screen slot openings. To determine this limiting size, a particle size analysis of the aquifer material must first be undertaken. About a cup of dry, thoroughly mixed aquifer material is passed through a standard set of sieves (Fig. 4.8) and the weight of the fractions retained on each sieve is recorded. These weights are then expressed as percentages of the total weight of sample and a graph is plotted of the cumulative percent of the sample retained on a given sieve and all the other sieves above it versus the size of the given sieve expressed in thousandths of an inch (Fig. 4.9). A smooth curve is drawn through the points on the graph. This curve shows at a glance how much of the material is smaller or larger than a given particle size. For example, the curve in Fig. 4.9 shows that 90 percent of the sample consists of sand grains larger than 0.010 inch or that 10 percent is smaller than this size. Expressed in another way, we may say that the 90 percent size of the sand is 0.010 inch.

Before describing the use of these sieve-analysis curves for the selection of screen slot openings it is desirable to point out another important use to which they are put. Reference here is to the use of the shape and location of the curve to determine the uniformity in size of the material and the classification of the material in such types as fine sands, coarse sands and gravels. For example, a narrowly spread, almost vertical type of curve indicates a uniform type of material. If such a curve occupies the left hand corner of the graph sheet (Fig. 4.10A) in the region of the small sieve sizes, then it represents a fine uniform sand. On the other hand, a curve widely spread across the graph sheet, as in Fig. 4.10D, indicates a sand and gravel mixture containing very little fine sand. An aquifer of such material would have a higher permeability and should be a much better producer of water than one containing the fine sand of Fig. 4.10A.

Examining Fig. 4.10D closely shows that removing all the material finer than the 40 percent size would leave only material coarser than 0.050 inch in

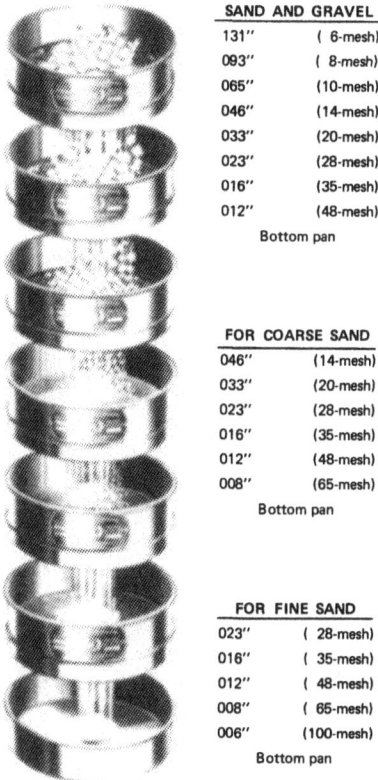

SAND AND GRAVEL

131"	(6-mesh)
093"	(8-mesh)
065"	(10-mesh)
046"	(14-mesh)
033"	(20-mesh)
023"	(28-mesh)
016"	(35-mesh)
012"	(48-mesh)
Bottom pan	

FOR COARSE SAND

046"	(14-mesh)
033"	(20-mesh)
023"	(28-mesh)
016"	(35-mesh)
012"	(48-mesh)
008"	(65-mesh)
Bottom pan	

FOR FINE SAND

023"	(28-mesh)
016"	(35-mesh)
012"	(48-mesh)
008"	(65-mesh)
006"	(100-mesh)
Bottom pan	

Fig. 4.8 RECOMMENDED SETS OF STANDARD SIEVES FOR ANALYZING SAMPLES OF WATER-BEARING SAND OR GRAVEL.

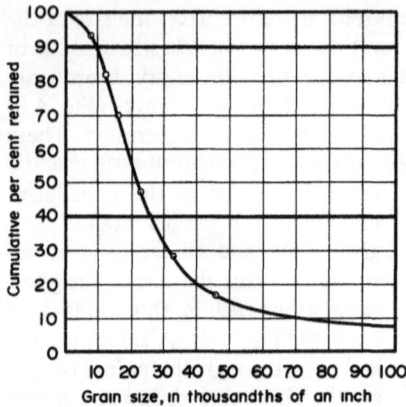

Size of Sieve Opening	Cumulative Weights Retained	Cumulative Per Cent Retained
0 046"	65 grams	17%
0 033"	106 grams	28%
0.023"	179 grams	47%
0 016"	266 grams	70%
0.012"	312 grams	82%
0 008"	357 grams	94%
Pan	380 grams	100%

Original weight = 382 grams

Fig. 4.9 TYPICAL SIEVE-ANALYSIS CURVE SHOWS DISTRIBUTION OF GRAIN SIZES IN PER CENT BY WEIGHT.

the formation. This relatively coarse material would have large pore spaces through which flow would be relatively free. A well constructed in aquifer material of this type with a screen carrying 0.050-inch slot openings or a No. 50 slot screen would have a high efficiency after proper development to remove the fine material.

Generally, well slot openings are designed to retain from 30 to 50 percent of the formation material depending upon the aquifer conditions. The selection should tend toward the higher value for fine, uniform sands containing corrosive waters and toward the lower value for coarse sand and gravel formations. For example, the 40 percent size is recommended for a fine, uniform sand if the water is non-corrosive. If the water were corrosive, however, this would cause a gradual enlarging of the slot openings with time and a resulting steady flow of sand into the well. The designer must be more conservative under such circumstances and select the smaller opening that would be given by the use of the 50 percent size. In a coarse sand and gravel formation, however, the enlarging of the selected slot opening by a few thousandths of an inch would not create a perpetual sanding problem and the 30 percent size may be chosen for the slot opening.

The selection of a 30 percent size of opening means that 70 percent of the formation in the vicinity of the well will be removed in the developing process. Similarly, 60 percent of the formation is removed with a 40 percent size of slot opening. Selecting the 30 percent size as against the 50 percent size means that more material is removed, thus causing the development of a larger zone in the material surrounding the screen. This usually increases the specific capacity of the well and hence its efficiency in sufficient proportion to offset the extra cost of development. This is only permissible if the formation conditions are such as to indicate the use of the larger 30 percent size of slot opening. A more conservative selection of slot size is recommended whenever there is doubt about the reliability of the samples provided for analysis.

44

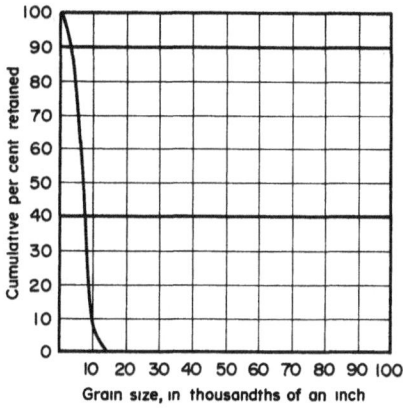

A. Fine, uniform sand that yields water at limited rates.

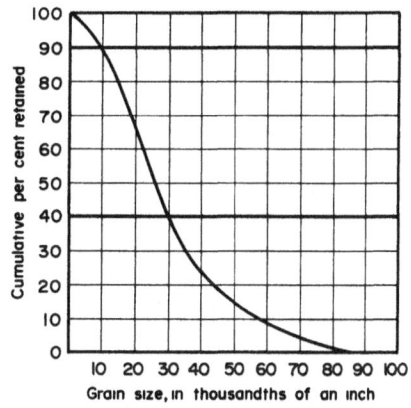

B. Medium and coarse sand mixture with good permeability.

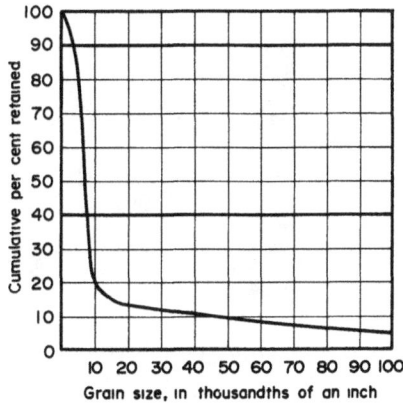

C. Fine sand with 10 to 20 percent coarse particles.

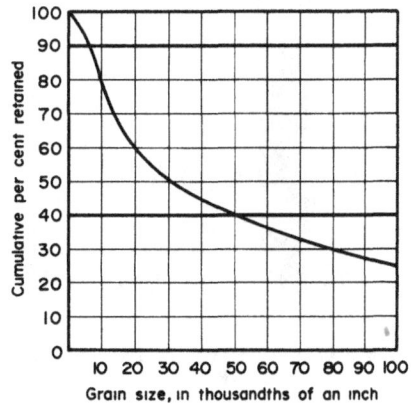

D. Sand and gravel mixture with good permeability.

Fig. 4.10 TYPICAL SIEVE-ANALYSIS CURVES FOR WATER-BEARING SANDS AND GRAVELS.

45

Most geologic formations are stratified, having layers of varying particle size distribution. In such cases, slot size openings should be selected for different sections of screen to suit the particle size distribution of the different strata. Two more rules should be followed in aquifers where a fine sand overlies coarse material.

1. The screen with the slot size designed for the finer material should be extended at least 2 feet into the coarse material.
2. The slot size of the screen designed for the coarse material should never be greater than twice the slot size for the overlying finer material.

These rules are aimed at reducing the possibility of the well perpetually producing sand from the fine upper layer. Fig. 4.11 illustrates how this possibility may arise. It should also be remembered that depths to formation changes are not always accurately measured and it is not always possible to set screens at the exact levels intended. The observation of these rules then assumes greater importance.

The method of selecting screen slot openings so far outlined assumes

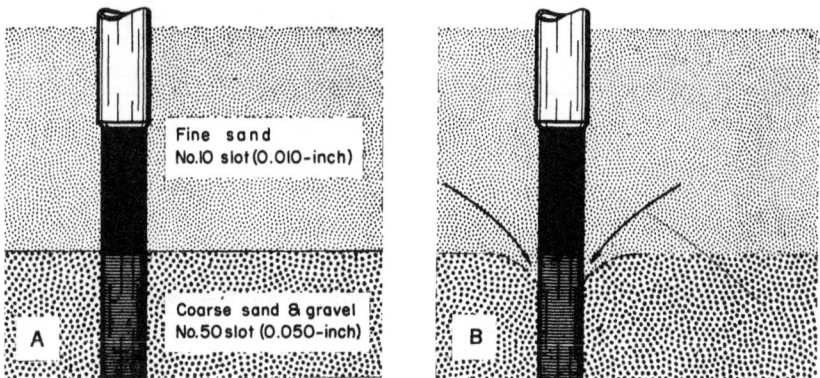

Fig. 4.11 SEQUENCE ILLUSTRATES POSSIBILITY OF FINE SAND ENTERING UP-PER PART OF LOWER SECTION OF SCREEN AFTER DEVELOPMENT OF WELL IF THE LARGER OPENINGS OF THIS LOWER SECTION OF SCREEN EXTEND TO THE TOP OF THE COARSE MATERIAL.

conditions that make it practicable to order well screens after doing sieve analyses of formation materials. In many countries and in the remote parts of some others this procedure would result in costly delays while awaiting an imported screen. The designer of small wells under such conditions would be justified in selecting a slot opening(s) based upon previous experience with existing wells in the same aquifer even before drilling operations begin. It would also be advisable to select a standard size of slot opening for a multiple-well program in the same aquifer in order to benefit from the resulting reduced costs and time saving. This may, however, entail gravel packing of some of the wells to prevent them from producing fine sand. The efficiency of other wells may be less than optimum. This, however, is not a

46

prime concern in small wells. Generally, the benefits of standardization of slot openings of small wells under the above stated conditions would offset the disadvantages.

The *entrance velocity* is determined by dividing the expected or desired yield of the well expressed in cubic feet per second by the total area of the screen openings expressed in square feet.

The total area of screen openings is the area of openings provided per foot of screen multiplied by the selected length of screen expressed in feet. Most manufacturers provide tables showing the open area per foot of screen for each size of screen diameter and for various widths of slot openings. Table 4.1 is an example of one of these. From this table it is seen that a No. 40 slot, 3-inch diameter telescope-size screen of this type contains 42 square inches of open area per foot of screen length. A 10-foot length of such a screen would, therefore, contain 420 square inches of total open area.

The design standard for the entrance velocity is chosen such that the friction losses in the screen openings will be negligible and the rate of incrustation and corrosion will be minimum. Laboratory tests and field

TABLE 4.1

INTAKE AREAS FOR SELECTED WIDTHS OF SLOT OPENINGS, (Square Inches per Lineal Foot of Screen).

Nominal Screen Size	Actual OD of Screen	Slot No. 10 (0.010") (0.25 mm)	Slot No. 20 (0.020") (0.50 mm)	Slot No. 40 (0.040") (1.00 mm)	Slot No. 60 (0.060") (1.50 mm)
2" TS	1-3/4"	10	16	26	32
1½" PS	2-3/8"	13	22	36	45
2" PS	2-5/8"	14	25	41	50
3" TS	2-3/4"	15	26	42	52
2½" PS	3-1/8"	17	30	48	59
3" PS	3-5/8"	20	34	54	68
4" TS	3-3/4"	21	35	56	71
4" PS	4-5/8"	25	44	68	86

(Courtesy UOP-Johnson Division, Universal Oil Products Company, St. Paul, Minnesota)

Notes: TS means telescope-size well screen
PS means pipe-size well screen

experience have shown that these objectives are achieved if the screen entrance velocity is equal to or less than 0.1 ft per sec. The screen length preferably, or the diameter as is practicable, should be increased if this velocity is greater than 0.1 ft per sec. On the other hand, if the entrance velocity is appreciably less than 0.1 ft per sec — say 0.05 ft per sec — the screen length may be reduced until the entrance velocity more nearly approaches the standard of 0.1 ft per sec.

SELECTION OF CASING AND SCREEN MATERIALS

The choice of materials that go into the construction of a well is a very important aspect of water well design. A well constructed of materials with little or no resistance to corrosion can be destroyed beyond usefulness by a highly corrosive water within a few months of completion. This will be the case no matter how excellent the other aspects of design. A poor selection of materials can also result in collapse of the well due to inadequate strength. The above are factors which have considerable influence on what is called the useful life of a well. In addition to these influences, the selection of materials also has considerable bearing on the cost of a well. The corrosion resistant metals, for example, are much more costly than ordinary steel. The choice of a suitable metal or the provision of a greater thickness of the same metal to meet strength requirements invariably results in higher costs. These considerations, therefore, indicate that the designer must exercise great care in the selection of materials for a well.

The designer usually makes his decision on the choice of materials after considering three main factors. These are *water quality, strength requirements* and *cost.*

Water Quality

Water quality, in this context, refers primarily to the *mineral content* of the water that will be produced by the well. Its effects on metal may be of two basic types. It may cause *corrosion* or *incrustation.* Some waters cause both corrosion and incrustation. Chemical analyses of water samples can indicate to the skilled interpreter whether a water is likely to be corrosive, incrusting, or both. Unless knowledge is already available on the nature of the water in the aquifer, it would be wise to seek the advice of a chemist with relevant experience before selecting materials for use in a well.

Corrosion is a process which results in the destruction of metals. Corrosive waters are usually acid and may contain relatively high concentrations of dissolved oxygen which is often necessary for and increases the rate of corrosion. High concentrations of carbon dioxide, total dissolved solids and hydrogen sulfide with its characteristic odor of rotten eggs are other indications of a likely corrosive water.

Besides water quality, there are other factors such as *velocity of flow* and *dissimilarity of metals* which contribute to the corrosion process. The greater the velocity of flow, the greater is the removal of the protective corrosion end products from the surface of the metal and hence the exposure of that surface to further corrosion. This is another important reason for keeping the velocity through screen openings within acceptable limits. The use of two or more different types of metals such as stainless steel and ordinary steel, or steel and brass or bronze should be avoided whenever possible. Corrosion is usually greatest at the points of contact or closest proximity of the metals.

Corrosion may occur in well screens as well as casings. It can be more critical in screens because it can reach damaging proportions much earlier than in casings. This is because only a small enlargement of the screen openings is required for the entry of sand through the screen, while the full thickness of the casing metal must be penetrated for failure of a well through corrosion of the casing. This is, however, no reason for ignoring the effect of corrosion in casings. Casing failure by corrosion equally ruins a well as does failure of the screen. It can cause the introduction of clay and polluted or

48

otherwise unsatisfactory water into the well. Corrosive well waters have been observed to destroy steel casings in less than 6 months in Guyana, thus ruining many wells.

Ordinary steel and iron are not corrosion resistant. There are, however, a number of metal alloys available with varying degrees of corrosion resistance. Among these are the stainless steels which combine nickel and chromium with steel and also the various copper-based alloys such as brass and bronze which combine traces of silicon, zinc and manganese with copper. Manufacturers, supplied with water analyses, can be expected to provide advice on the type of metal or metal alloys to be used.

Plastic pipe of the polyvinyl chloride (pvc) type is an attractive alternative to the use of metals in small wells, particularly under corrosive conditions. It combines corrosion resistance with adequate strength and economy.

Incrustation, unlike corrosion, results not in the destruction of metal, but in the deposition of minerals on it and in the aquifer immediately around a well. Physical and chemical changes in the water in the well and the adjacent formation cause dissolved minerals to change to their insoluble states and settle out as deposits. These deposits cause the blocking of screen openings and the formation pore spaces immediately around the screen with a resulting reduction in the yield of the well.

Incrusting waters are usually alkaline or the opposite to corrosive waters, which are acid. Excessive carbonate hardness is a common source of incrustation in wells. Scale deposits of calcium carbonate (lime scale) occur in pipes carrying hard waters. Iron and manganese, to a lesser extent, are other common sources of incrustation in wells. Iron causes characteristic reddish-brown deposits while those of manganese are black.

Often associated with iron-containing ground waters are iron bacteria. These minute living organisms are non-injurious to health, but, while aiding the deposition of iron, produce accumulations of slimy, jelly-like material which block well screen openings and aquifer pore spaces.

Strong solutions of hydrochloric acid are often used in treatment processes for the removal of all the above-mentioned incrusting deposits. The corrosive effect of this acid treatment, which must be repeated as the need arises, makes it necessary to use screens made of corrosion-resistant materials. Unplasticized polyvinyl chloride pipe would also withstand such treatment. Further discussion on rehabilitating incrusted wells is presented in Chapter 7.

Strength Requirements

Strength requirements are important in both casing and screens but are generally of more concern in screens. Screens must be strong enough to withstand the external radial pressures that could cause their collapse as well as the vertical loading due to the weight of the casing above them.

Some metals have greater strength characteristics than others. Stainless steel, for example, can be twice as strong as some copper alloys. Screens and casings of adequate strength can be made from any of the metals and alloys commonly used in well construction. Manufacturers usually specify conditions under which their pipes and screens can be satisfactorily used. It is often helpful to consult with them on the selection of suitable materials for use in a well.

Cost

Cost considerations may often be the deciding factor in the selection of construction materials used in small wells. The situation may arise, for instance, where stainless steel would be the most suitable material for use, combining corrosion resistance with excellent strength and a long, useful life. However, its cost may cause the designer to recommend the use of some other less suitable material after weighing the benefits of extra useful life against lower initial cost, the cost of replacement at a later date and the owner's financial capacity.

Miscellaneous

Other miscellaneous factors also play important roles in the selection of casing and screen materials. Chief among these, with reference to small wells, would be site accessibility, ease of handling, availability, and on-site fabrication. In areas not accessible by motor vehicles and necessitating the use of air transportation, weight of materials could be the most decisive consideration. The lighter plastic-type materials would then gain preference over metals. Ease of handling, both for transportaiton and construction purposes, would also favor the use of plastic-type material.

The above are only some of the major considerations in the selection of materials. Solutions cannot be blindly transferred from one geographic area to another. Each set of conditions, and the advantages and disadvantages of each possible solution, must be carefully considered before making a final selection.

GRAVEL PACKING AND FORMATION STABILIZATION

Both gravel packing and formation stabilization are aids to the process of well development described earlier in this chapter. A further similarity is the addition of gravel in the case of gravel packing, and coarse sand or sand and gravel in the case of formation stabilization to the annular space between the screen and water-bearing formation. This, however, is where the similarities end. The differences between gravel packing and formation stabilization are indeed very fundamental and should be thoroughly grasped.

It will be recalled that the development process in a naturally developed well removes the finer material from the vicinity of the well screen, leaving a zone of coarser graded material around the well. This cannot be achieved in a formation consisting of a fine uniform sand due to the absence of any coarser material. The object of *gravel packing* a well is to artificially provide the graded gravel or coarser sand that is missing from the natural formation. A well treated in this manner is referred to as an artificially gravel-packed well to distinguish it from the naturally developed well.

Drilling by the rotary method through an unconsolidated water-bearing formation of necessity results in a hole somewhat larger than the outside diameter of the well screen. This provides the necessary clearance to permit the lowering of the screen to the bottom of the hole without interference.

The object of *formation stabilization* is to fill the annular space around the screen (possibly 2 inches and more in width) at least partially, to prevent the silt and clay materials above the aquifer from caving or slumping when the development work is started. By avoiding such caving, proper development of the well may be carried out with less time and effort. Note that the development process here is a natural one, with the graded coarse material coming from the aquifer itself and not from the added stabilizing material. The objectives of gravel packing and formation stabilization, therefore, provide the major difference between the two processes. These differences in objectives also form the basis for the differences in the design features of the two processes.

Gravel Packing

There are essentially two conditions in unconsolidated formations which tend to favor artificial gravel-pack construction.

The first of these, *fine uniform sand,* has already been mentioned. Such a sand would require a screen with very small slot openings and, even so, the development process would not be satisfactory because of the uniformity of the sand particles. Also, screens with very small slot openings have low percentages of open area because of the relative thickness of the metal wires that must be used to provide strength. By artificially gravel packing wells in such formations, screens with larger slot openings may be used and the improved development results in greater well efficiency. The use of artificial gravel-pack construction is recommended in formations where the screen slot opening, selected on the basis of a naturally developed well, is smaller than 0.010 inch (No. 10 slot).

Extensively laminated formations provide the second set of conditions for which gravel pack construction is recommended. This refers to those aquifers that consist of thin, alternating layers of fine, medium, and coarse sand. In such aquifers it is difficult to accurately determine the position and thickness of each individual layer and to choose the proper length of each section of a multiple-slot screen. The use of artificial gravel packing in such formations reduces the chances of error that would result from natural development.

Selection of gravel-pack material: The selection of the grading of gravel-pack material is usually based on the layer of finest material in an aquifer. The gravel-pack material should be such that (1) its 70 percent size is 4 to 6 times the 70 percent size of the material in the finest layer of the aquifer, and (2) its uniformity coefficient is less than 2.5, and the smaller the better. *Uniformity coefficient* is the number expressing the ratio of the 40 percent size of the material to its 90 percent size. It is well to recall here that the sizes refer to the percentage retained on a given sieve.

The first condition usually ensures that the gravel-pack material will not restrict the flow from the layers of coarsest material, the permeability of the pack being several times that of the coarsest stratum. The second condition ensures that the losses of pack material during the development work will be minimal. To achieve this goal, the screen openings are chosen so as to retain 90 percent or more of the gravel-pack material.

Gravel-pack material should consist of clean, well rounded, smooth grains. Quartz and other silica-based materials are preferable. Limestone and shale are undesirable in gravel-pack material.

Thickness of gravel-pack envelopes: Gravel-pack envelopes are usually 3 to 8 inches thick. This is not out of necessity as tests have shown that a fraction of an inch would satisfactorily retain and control the formation sand. The greater thicknesses are used in order to ensure that the well screen is completely surrounded by the gravel-pack material.

Formation Stabilization

The quantity of formation stabilizer should be sufficient to fill the annular space around the screen and casing to a level about 30 feet, or as much as is practicable, above the top of the screen. This would allow for settlement and losses of the material through the screen during development. If necessary, more material should be added as development proceeds to prevent its top level from falling below that of the screen. The settlement of the material is beneficial in eroding the mud wall formed in boreholes drilled by the rotary method, thus making well development much easier.

The typical concrete or mortar sand is widely used as a formation stabilizer. The aquifer conditions under which it is suitable range from those requiring a No. 20 (0.020-inch) to those of a No. 50 (0.050-inch) slot opening. A specially graded material is not necessary.

SANITARY PROTECTION

It has been stated in Chapter 2 that ground waters are generally of good sanitary quality and safe for drinking. Well design should be aimed at the extraction of this high quality water without contaminating it or making it in any way unsafe for human consumption. The penetration of a water-bearing formation by a well provides two main routes for possible contamination of the ground water. These are the open, top end of the casing and the annular space between the casing and the borehole. The designer must concern himself with the prevention of contamination through these two routes.

Upper Terminal

Well casing should extend at least 1 foot above the general level of the surrounding land surface. It should be surrounded at the ground surface by a 4-inch thick concrete slab extending at least 2 feet in all directions. The upper surface of this slab and its immediate surroundings should be gently sloping so as to drain water away from the well, as shown in Fig. 4.12. It is also good practice to place a drain around the outer edge of the slab and extend it to a discharge point at some distance from the well. A sanitary well seal should be provided at the top of the well to prevent the entrance of contaminated water or other objectionable material directly into the well. Examples of these are shown in Fig. 4.13.

Lower Terminal of the Casing

For artesian aquifers, the water-tight casing should be extended downwards into the impermeable formation (such as a clay) which caps the

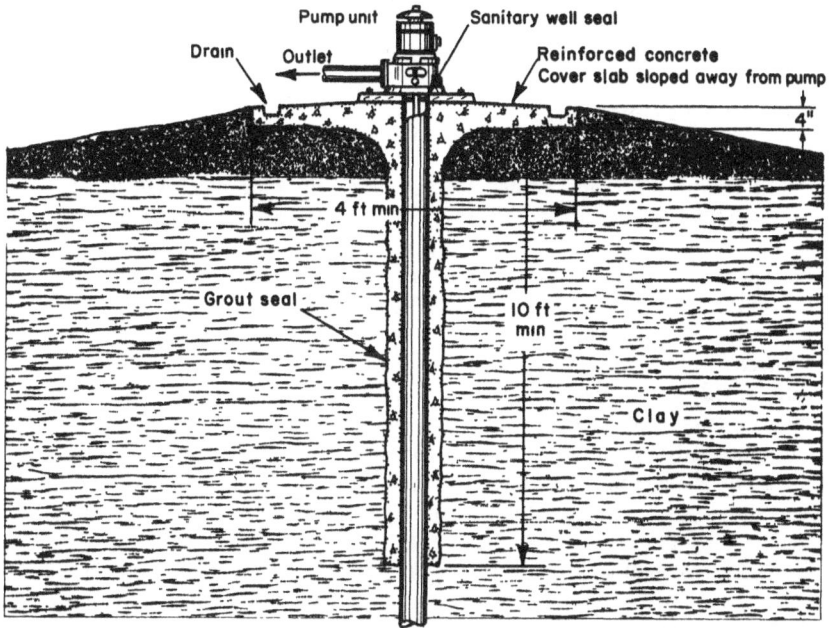

Fig. 4.12 SANITARY PROTECTION OF UPPER TERMINAL OF WELL.

aquifer. The purpose of this is to retain the artesian pressure of the aquifer by providing a seal against leakage from the aquifer up the outside of the casing. The borehole should not be extended into the artesian aquifer until the casing has been set and grouted.

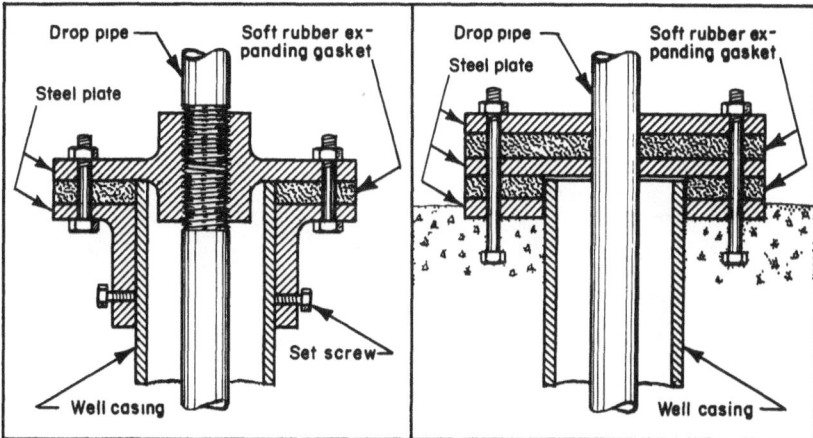

Fig. 4.13 SANITARY WELL SEALS. (Adapted from Fig. 7, *Manual of Individual Water Supply Systems*, Public Health Service Publication No. 24, 1962.)

In water-table aquifers the casing should be extended at least 5 feet below the lowest expected pumping level. This limiting distance should be increased to 10 feet where the pumping level is less than 25 feet from the surface.

The above are general rules which should be applied with some flexibility where geologic conditions so require.

Grouting and Sealing Casing

The drilled hole must of necessity be larger than the pipe used for the well casing. This results in the creation of an irregularly shaped annular space around the casing after it has been placed in position. It is important to fill this space in order to prevent the seepage of contaminated surface water down along the outside of the casing into the well and also to seal out water of unsuitable quality in strata above the desirable water-bearing formation.

In caving material, such as sand or sand and gravel, the annular space is soon filled as a result of caving. In such cases, therefore, no special arrangements need be made for filling the annular space. However, where the material overlying the water-bearing formation is of the non-caving type, such as clay or shale, then the annular space should be grouted with a cement or clay slurry to a minimum depth of 10 feet below the surface. Where the thickness of the clayey materials permit it, increasing the depth of grout to about 15 feet would provide added safety. The diameter of the drilled hole should be 3 to 6 inches larger than the permanent well casing to facilitate the placing of the grout. It is important to remove temporary casing when grouting rather than simply filling the space between the two casings as vertical seepage can readily occur down the outside of any unsealed casing. Methods of mixing and placing the grout are discussed in Chapter 5.

54

CHAPTER 5

WELL CONSTRUCTION

There are four basic operations involved in the construction of tubular wells. These are the drilling operation, casing installation, grouting of the casing when necessary and screen installation.

WELL DRILLING METHODS

The term well drilling methods is being used here to include all methods used in creating holes in the ground for well construction purposes. As such, it includes methods such as boring and driving which are not drilling methods in a pure sense. The classification is one of convenience in the absence of a better descriptive term. The limitations on well diameter (4 inches and less) exclude the dug well from consideration. The sections that follow describe the bored and driven, the percussion, hydraulic rotary and jet drilled wells.

Boring

Boring of small diameter wells is commonly undertaken with hand-turned earth augers, though power-operated augers are sometimes used. Two common types of hand augers are shown in Fig. 5.1. They each consist of a shaft

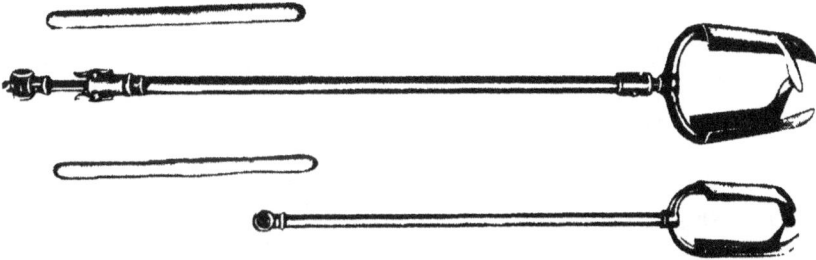

Fig. 5.1 HAND AUGERS. (From Fig. 6, *Wells*, Department of the Army Technical Manual TM5-297, 1957.)

with wooden handle at the top and a bit with curved blades at the bottom. The blades are usually of the fixed type, but augers with blades that are adaptable to different diameters are also available. Shafts are usually made up of 5-ft sections with easy latching couplings.

The hole is started by forcing the blades of the bit into the soil with a turning motion. Turning is continued until the auger bit is full of material. The auger is then lifted from the hole, emptied and returned to use. Shaft extensions are added as needed to bore to the desired depth. Wells shallower than 15 ft ordinarily require no other equipment than the auger. Deeper wells, however, require the use of a light tripod with a pulley at the top, or a raised platform, so that the auger shaft can be inserted and removed from the hole without disconnecting all shaft sections.

The spiral auger shown in Fig. 5.2 is used in place of the normal cutting bit to remove stones or boulders encountered during boring operations. When turned in a clockwise direction, the spiral twists around a stone so that it can be lifted to the surface.

The method is used in boring to depths of about 50 ft in clay, silt and sand formations not subject to caving. Boring in caving formations may be done by lowering casing to the bottom of the hole and boring ahead little by little while forcing the casing down.

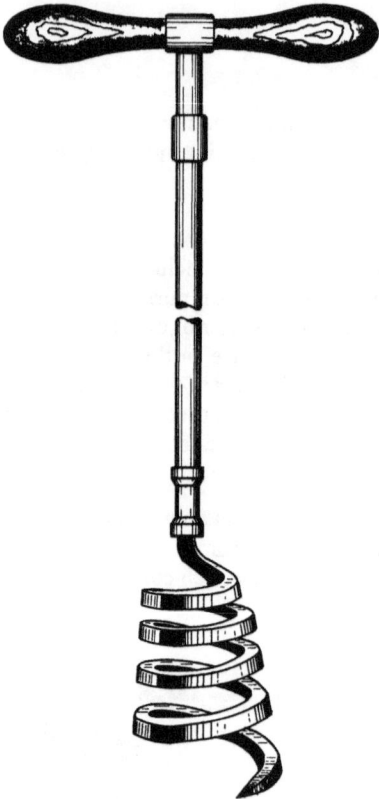

Fig. 5.2   SPIRAL AUGER.

Driving

Driven wells are constructed by driving into the ground a well point fitted to the lower end of tightly connected sections of pipe. The well point must be sunk to some depth within the aquifer and below the water table. The riser pipe above the well point functions as the well casing.

Equipment used includes a drive hammer, drive cap to protect the top end of the riser pipe during driving, tripod, pulley and strong rope with or without a winch. A light drilling rig may be used instead of the tripod assembly. Well points can be driven either by hand methods or with the aid of machines. Fig. 5.3 shows the assembly for a purely hand-driven method. The drive-block assemblies commonly operated by a drilling rig or by hand with the aid of a tripod and tackle are shown in Fig. 5.4.

Whatever the method of driving, a starting hole is first made by boring or digging to a depth of about 2 feet or more. As driving is generally easier in a saturated formation, the starting hole should be made deep enough to penetrate the water table if the latter is sufficiently shallow. The starting hole should be vertical and slightly larger in diameter than the well point. The well

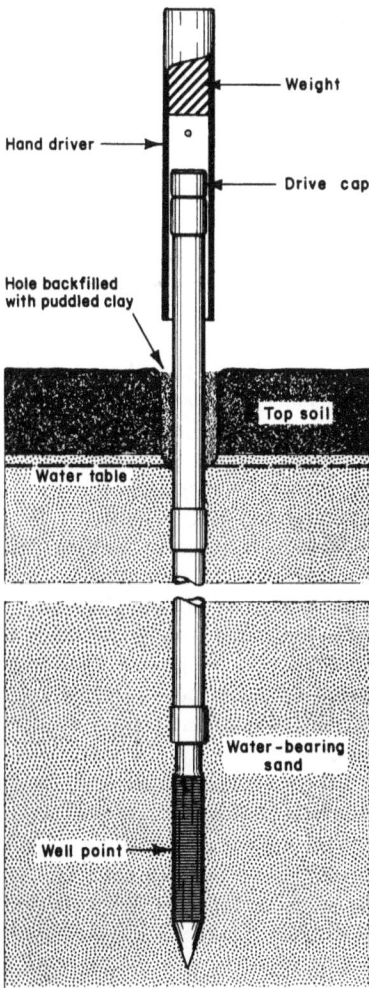

Fig. 5.3 SIMPLE TOOL FOR DRIV-
ING WELL POINTS TO
DEPTHS OF 15 TO 30 FT.

point is inserted into this hole and driven to the desired depth, 5-ft lengths of riser pipe being added as necessary. Pipe couplings should have recessed ends and tapered threads to provide stronger connections than ordinary plumbing couplings. The pipe and coupling threads should be coated with pipe thread compound to provide airtight joints. The well-point assembly should be guided as vertically as possible and the driving tool, when suspended, should be hung directly over the center of the well. The weight of the driving tool may range from 75 to 300 pounds. Heavier tools require the use of a power hoist or light drilling rig. The spudding action of a cable-tool drilling machine (Fig. 5.14) is well suited for rapid well point driving. Slack joints should be periodically tightened by turning the pipe lightly with a wrench. Violent twisting of the pipe makes driving no easier and can result in damage to the well point. This must, therefore, be avoided.

Driven wells can be installed only in unconsolidated formations relatively free of cobbles and boulders. Hand driving can be undertaken to depths up to about 30 feet; machine driving can achieve depths of 50 feet and greater.

Jetting

The jetting method of well drilling uses the force of a high velocity stream or jet of fluid to cut a hole into the ground. The jet of fluid loosens the subsurface materials and transports them upward and out of the hole. The rate of cutting can be improved with the use of a drill bit (Fig. 5.5) which can be rotated as well as moved in an up-and-down chopping manner.

The fluid circulation system is similar to that of conventional rotary drilling described later in this chapter. Indeed the equipment can be identical with that used for rotary drilling, with the exception of the drill bit. Simple equipment for jet drilling is shown in Fig. 5:6. A tripod made of 2-inch

57

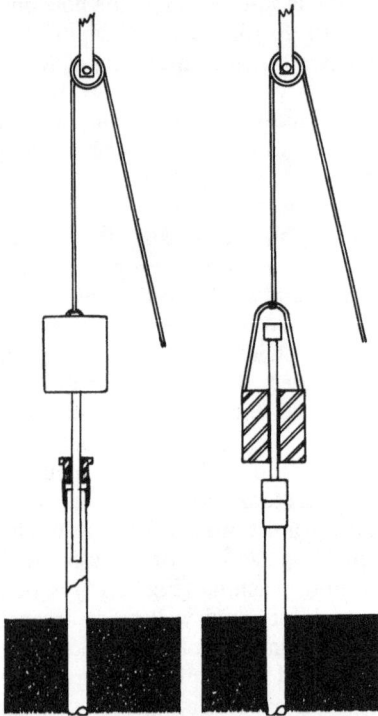

Fig. 5.4 DRIVE-BLOCK ASSEMBLIES FOR DRIVING WELL POINTS.

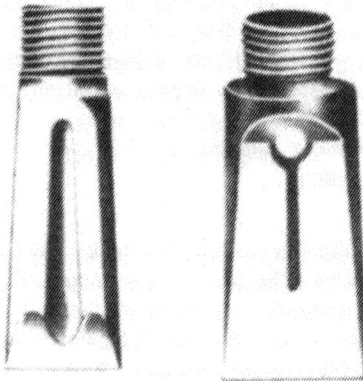

Fig. 5.5 BITS FOR JET DRILLING. (From Fig. 17, *Wells*, Department of the Army Technical Manual TM5-297, 1957.)

galvanized iron pipe is used to suspend the galvanized iron drill pipe and the bit by means of a U-hook (at the apex of the tripod), single-pulley block and manila rope. A pump having a capacity of approximately 150 gallons per minute at a pressure of 50 to 70 pounds per square inch is used to force the drilling fluid through suitable hose and a small swivel on through the drill pipe and bit. The fluid, on emerging from the drilled hole, travels in a narrow ditch to a settling pit where the drilled materials (cuttings) settle out and then to a storage pit where it is again picked up by the pump and recirculated. The important features of settling and storage pits are described in the later section of this chapter dealing with hydraulic rotary drilling. A piston-type reciprocating pump would be preferred to a centrifugal one because of the greater maintenance required by the latter as a result of leaking seals and worn impellers and other moving parts.

The spudding percussion action can be imparted to the bit either by means of a hoist or by workmen alternately pulling and quickly releasing the free end of the manila rope on the other side of the block from the swivel. This may be done while other workmen rotate the drill pipe. The drilling fluid may be and is very often plain water. Depths of the order of 50 feet may be achieved in some formations using water as drilling fluid without undue caving. When caving does occur, then a drilling mud as described in the later section on hydraulic rotary drilling should be used.

The jetting method is particular-

Fig. 5.6 SIMPLE EQUIPMENT FOR JET OR ROTARY DRILLING.

ly successful in sandy formations. Under these conditions a high rate of penetration is achieved. Hard clays and boulders do present problems.

Hydraulic Percussion

The hydraulic percussion method uses a similar string of drill pipe to that of the jetting method. The bit is also similar except for the ball check valve placed between the bit and the lower end of the drill pipe. Water is introduced continuously into the borehole outside of the drill pipe. A reciprocating, up-and-down motion applied to the drill pipe forces water with suspended cuttings through the check valve and into the drill pipe on the down stroke, trapping it as the valve closes on the up stroke. Continuous reciprocating motion produces a pumping action, lifting the fluid and cuttings to the top of the drill pipe where they are discharged into a settling tank. The cycle of circulation is then complete. Casing is usually driven as drilling proceeds.

The method uses a minimum of equipment and provides accurate samples of formations penetrated. It is well suited for use in clay and sand formations that are relatively free of cobbles or boulders.

Sludger

The sludger method is the name given to a forerunner of the hydraulic percussion method described in the previous section. It is accomplished entirely with hand tools, makes use of locally available materials, such as

59

bamboo for scaffolding, and is particularly suited to use in inaccessible areas where labor is plentiful and cheap. The first description of the method is believed to have come from East Pakistan where it has been used extensively.

In the sludger method, as used in East Pakistan, scaffolding is erected as shown in Fig. 5.7. The reciprocating, up-and-down motion of the drill pipe is provided by means of the manually operated bamboo lever to which the drill pipe is fastened with a chain. A sharpened coupling is used as a bit at the lower end of the drill pipe. The man shown seated on the scaffolding uses his hand to perform the functions of the check valve as used in the hydraulic percussion method, though, in this case at the top instead of the bottom of the drill pipe. A pit, approximately 3 feet square and 2 feet deep, around the drill pipe, is filled with water which enters the borehole as drilling progresses. On the upstroke of the drill pipe its top end is covered by the hand. The hand is removed on the downstroke (Fig. 5.8), thus allowing some of the fluid and cuttings sucked into the bottom of the drill pipe to rise and overflow. Continuous repetition of the process causes the penetration of the drill pipe into the formation and creates a similar pumping action to that of the hydraulic percussion method. New lengths of drill pipe are added as necessary. The workman whose hand operates as the flap valve changes position up and down the scaffolding in accordance with the position of the top of the drill pipe. Water is added to the pit around the drill pipe as the level drops. When the hole has been drilled to the desired depth, the drill pipe is extracted in sections, care being taken to prevent caving of the borehole. The screen and casing are then lowered into position.

Wells up to 250 feet deep have been drilled by this method in fine or sandy formations. Reasonably accurate formation samples can be obtained during drilling. Costs are confined to labor and the cost of pipe, and can therefore, be very low. The method requires no great operating skills.

Fig. 5.7 BAMBOO SCAFFOLDING, PIVOT AND LEVER USED IN DRILLING BY THE SLUDGER METHOD. (From "Jetting S m a l l Tubewells By Hand," *Water Supply and Sanitation in Developing Countries,* AID-UNC/IPSED Item No. 15, June 1967.)

Hydraulic Rotary

Hydraulic rotary drilling combines the use of a rotating bit for cutting the borehole with that of continuously circulated drilling fluid for removal of the cuttings. The basic parts of a conventional rotary drilling machine or rig are a derrick or mast and hoist; a power operated revolving table that rotates the

60

Fig. 5.8 MAN ON SCAFFOLDING RAISES HAND OFF PIPE ALLOWING DRILL FLUID AND CUTTINGS TO ESCAPE. (From "Jetting Small Tube-wells By Hand," *Water Supply and Sanitation in Developing Countries,* AID-UNC/IPSED Item No. 15, June, 1957.)

drill stem and drill bit below it; a pump for forcing drilling fluid via a length of hose and a swivel on through the drill stem and bit; and a power unit or engine. The drill stem is in effect a long tubular shaft consisting of three parts: the kelly; as many lengths of drill pipe as required by the drilling depth; and one or more lengths of drill collar.

The kelly or the uppermost section of the drill stem is made a few feet longer and of greater wall thickness than a length of drill pipe. Its outer shape is usually square (sometimes six-sided or round with lengthwise grooves), fitting into a similarly shaped opening in the rotary table such that the kelly can be freely moved up or down in the opening even while being rotated. At the top end of the kelly is the swivel which is suspended from the hook of a traveling hoist block.

Below the kelly are the drill pipes, usually in joints about 20 feet long. Extra heavy lengths of drill pipe called drill collars are connected immediately above the bit. These add weight to the lower end of the drill stem and so help the bit to cut a straight, vertical hole.

The bits best suited to use in unconsolidated clay and sand formations are drag bits of either the fishtail or three-way design (Fig. 5.9). Drag bits have short blades forged to thin cutting edges and faced with hard-surfacing metal. The body of the bit is hollow and carries outlet holes or nozzles which direct the fluid flow toward the center of each cutting edge. This flow cleans and cools the blades as drilling progresses. The three-way bit performs smoother and faster than the fishtail bit in irregular and semi-consolidated formations and has less tendency to be deflected. It cuts a little slower than the fishtail bit, however, in truly unconsolidated clay and sand formations.

Coarse gravel formations and those containing boulders may require the use of roller-type bits shown in Fig. 5.10. These bits exert a crushing and chipping action as they are rotated, thus cutting harder formations effectively. Each roller is provided with a nozzle serving the same purpose with respect to the rollers as those on the drag bits with respect to their blades.

The pump forces the drilling fluid through the hose, swivel, rotating drill stem and bit into the drilled hole. The drill fluid, as it flows up and out of the drilled hole, lifts the cuttings to the ground surface. At the surface the fluid flows in a suitable ditch to a settling pit where the cuttings settle out. From here it overflows to a storage pit where it is again picked up by the pump and recirculated. The settling pit should be of volume equal to at least three times

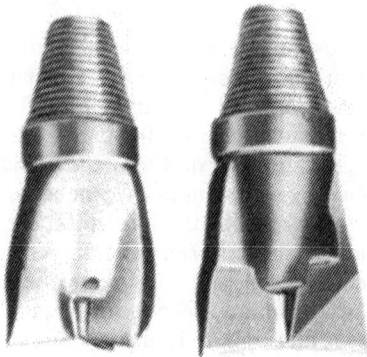

Fig. 5.9 ROTARY DRILL BITS. (From Fig. 41, *Wells*, Department of the Army Technical Manual TM5-297, 1957.)

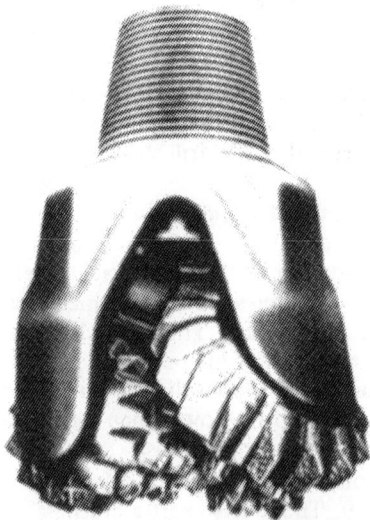

Fig. 5.10 ROLLER-TYPE ROTARY DRILL BIT. (From Reed Drilling Tools, Houston, Texas.)

the volume of the hole being drilled. It should be relatively shallow (a depth of 2 feet to 3 feet usually proving satisfactory) and about twice as long in the direction of flow as it is wide and deep. In accordance with the above rules a settling pit 6 feet long, 3 feet wide and 3 feet deep would be suitable for the drilling of 4-inch wells (hole diameter of 6 inches) 100 feet in depth. A system of baffles may also be used to provide extra travel time in the pit and thus improve the settling.

The storage pit is intended mainly to provide enough volume from which to pump. A pit 3 feet square and 3 feet deep would be satisfactory. It may either be combined with the settling pit to form a single, larger pit or separated from the settling pit by a connecting ditch. Drill hole cuttings should be periodically removed from the pits and ditches as is necessary.

The drilling fluid performs other important functions in the drilled hole besides those already mentioned. These are discussed later in this chapter.

Fig. 5.11 shows a number of the component parts of a rotary drilling rig. The chain pulldowns shown are used mainly for applying greater downward force to the drill pipe and bit but are not normally required for the drilling of small wells in unconsolidated formations.

Rotary drilling equipment for small diameter shallow wells can be much simpler and less sophisticated than that just described. The truck, trailer or skid mounted derrick or mast can be substituted by a tripod made of 2-inch or 3-inch galvanized iron pipe. A small suitable swivel can be suspended by rope through a single-pulley block from a U-hook fixed by a pin at the apex of the tripod. Drill pipe and bits both made from galvanized iron

Fig. 5.11 ROTARY DRILLING RIG. (From The Winter Weiss Company, Denver, Colorado.)

pipe, a suitable pump, length of hose and hoist then complete the requirements. One or two men can use chain tongs to rotate the drill pipe. With the exception of the drilling bit, this equipment can be identical with that described for the jetting method and shown in Fig. 5.6. This simple drilling equipment is light, manageable and easily transported to areas that are inaccessible to larger rigs.

Fig. 5.12 shows a fishtail-type drilling bit made from a short length of galvanized pipe. This type of bit has been successfully used by drillers of the Ministry of Works and Hydraulics, Guyana, South America for the drilling of shallow, small diameter wells in deltaic clay and sand formations. The bits are inexpensive and easy to make. In addition, they provide some means of use for the ever-available short ends of galvanized pipe.

Holes drilled by the rotary method in unconsolidated formations generally tend to collapse unless the properties of the drilling fluid (drilling mud) are such as to provide adequate support for the wall of the hole. Drilling muds are usually viscous mixtures of water, natural or commercial clays such as bentonite and sometimes other special purpose materials. The weight of this muddy fluid in the hole must be such as to provide enough pressure to exceed the earth pressure and any artesian pressure in the aquifer tending to cause

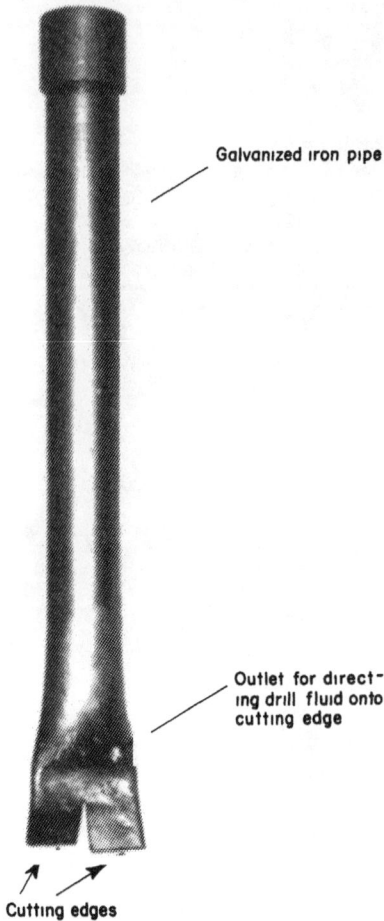

Galvanized iron pipe

Outlet for directing drill fluid onto cutting edge

Cutting edges

Fig. 5.12 FISHTAIL-TYPE ROTARY DRILL BIT MADE FROM GALVANIZED IRON PIPE.

collapse. In addition, the drilling mud forms a mud cake or rubbery sort of lining on the wall of the borehole. This mud cake holds the loose particles of the formation in place, protects the wall from being eroded by the upward stream of fluid and seals the wall to prevent loss of fluid into permeable formations such as sands and gravels. Drillers must be careful not to increase the pumping rate to the point where it causes destruction of the mud cake and caving of the hole.

The drilling fluid must also be such that the clay doesn't settle out of the mixture when pumping ceases but remains somewhat elastic, thus keeping the cuttings in suspension. All natural clays do not exhibit this property, known as *gelling*. Bentonite clays do exhibit satisfactory gel strength and are added to natural clays to improve their gel properties to desired levels.

The driller must also use his good judgement in arriving at a suitable fluid thickness. Too thin a fluid results in caving of the hole and loss of fluid into permeable formations. On the other hand, fluid that is too thick can cause difficulty in pumping. The drilling fluid should be no thicker than is necessary to maintain a stable hole and satisfactory removal of cuttings from it. The experienced driller often can adjust his fluid mixture to a satisfactory level by inspection. There are, however, two aids which a driller can use in the field to check the fluid characteristics and exert the necessary control. These are a balance for determining the density of the mud and a Marsh funnel for determining its viscosity Both of these are shown in Fig. 5.13. The balance has a cup at one end and a sliding weight on the other portion of its beam. The weight is moved until it balances the cup filled with the drilling fluid. The density of the fluid is then read from the balance arm which is calibrated in pounds per gallon. For most water-well drilling, a fluid with density of about 9 pounds per gallon is usually satisfactory.

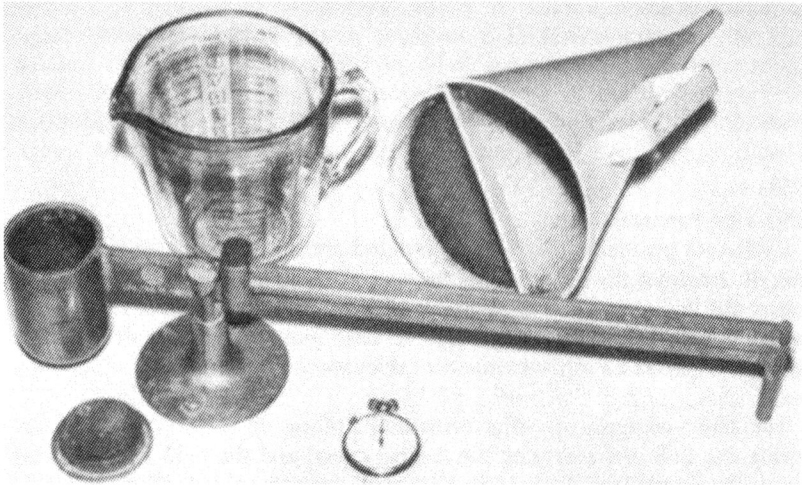

Fig. 5.13 BALANCE FOR DETERMINING MUD WEIGHT, STOP WATCH, MARSH FUNNEL FOR MEASURING MUD VISCOSITY AND 1-QUART MEASURING CUP.

To determine the viscosity, the lower end of the Marsh funnel is blocked by a finger while filling the funnel to the proper level (a volume of 1,500 cu cm). The finger is then removed to allow the fluid to discharge from the funnel. The time, in seconds, required to drain 1000 cu cm (approximately 1 quart) of the fluid is defined as the Marsh-funnel viscosity expressed in seconds.

A good drilling mud of density 9 pounds per gallon would have a Marsh-funnel viscosity in the range of 30 to 40 seconds. Sand picked up by the drilling mud from cuttings has the effect of increasing the density while reducing the Marsh-funnel viscosity. In contrast, native clays can be expected to increase both the density and viscosity of the fluid. Water and/or clay should be added periodically to the drilling mud as is necessary to keep the density and viscosity within the above limits.

Drilling by the hydraulic rotary method usually penetrates unconsolidated formations faster than is achievable by any other method. This can result in appreciable savings in time and cost, both of which can be major considerations in a well construction program. Since the borehole need not be cased until drilling is complete, the hole can be abandoned if necessary without the trouble of pulling or leaving the string of casing behind. A third advantage is the greater ease with which artificially gravel-packed wells can be constructed in unconsolidated formations, particularly when two or more zones are to be developed.

The hydraulic rotary method also has some disadvantages. The accurate sampling and logging of the formations penetrated can be difficult for the inexperienced driller because of the differential rate of transport of the cuttings out of the borehole. The need for proper drilling mud control also requires considerable experience on the part of the rotary driller. The training of rotary drillers can be more time consuming and difficult. Despite these disadvantages the method finds considerable application in the construction of wells in all types of formations and particularly unconsolidated formations.

Cable-Tool Percussion

Cable-tool percussion is one of the oldest methods used in well construction. It employs the principle of a free-falling heavy bit delivering blows against the bottom of a hole and thus penetrating into the ground. Cuttings are periodically removed by a bailer or sand pump. Tools for drilling and bailing are carried on separate lines or cables spooled on independent hoisting drums.

The basic components of a cable-tool drilling rig are a power unit for driving the bull reel (carrying the drilling cable) and the sand reel (carrying the bailing cable), and a spudding beam for imparting the drilling motion to the drill tools, all mounted on a frame which carries a derrick or mast of suitable height for the use of a string of drilling tools. Fig. 5.14 shows a cable-tool drilling rig on location.

Fig. 5.14 STAR 71 CABLE-TOOL DRILLING RIG.

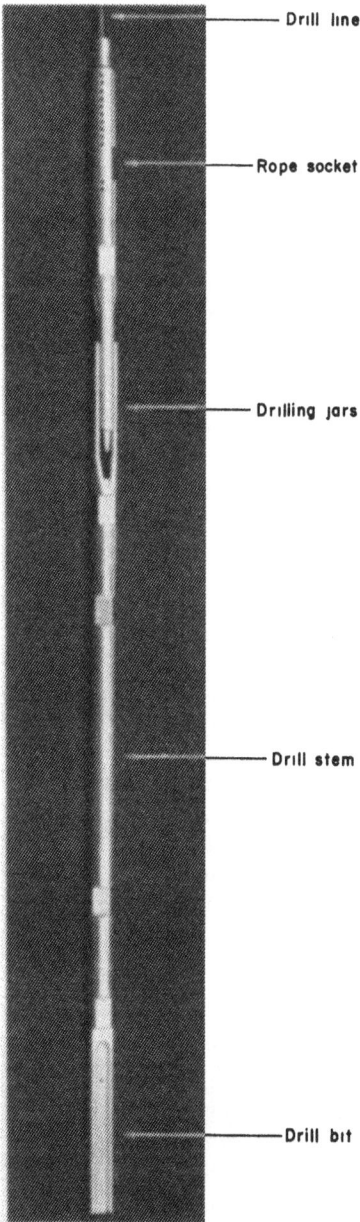

Fig. 5.15 COMPONENTS OF A STRING OF DRILL TOOLS FOR CABLE-TOOL PERCUSSION METHOD. (From Acme Fishing Tool Company, Parkersburg, West Virginia.)

Four items comprise a full string of drilling tools. These are the drill bit, drill stem, drilling jars and rope socket (Fig. 5.15). The chisel-shaped drill bit is used to loosen unconsolidated rock materials and with its reciprocating action mix these materials into a slurry which is later removed by bailing. When drilling in dry formations water must be added to the hole to form the slurry. The water course on the bit permits the movement of the slurry relative to the bit and, therefore, aids in the free-falling reciprocating motion of the bit. The drill stem immediately above the bit merely gives additional weight to the bit and added length to the string of tools to help maintain a straight hole.

The jars consist of a pair of linked steel bars which can be moved in a vertical direction relative to each other. The gap or stroke of the drilling jars is 6 to 9 inches. Jars are used to provide upward blows when necessary to free a string of tools stuck or wedged in the drill hole. Drilling jars are to be differentiated from similarly constructed fishing jars which have a stroke of 18 to 36 inches and are used in fishing or recovering tools which have come loose from the string of drilling tools in the hole.

The rope socket connects the string of tools to the cable. Its construction is such as to provide a slight clockwise rotation of the drilling tools relative to the cable. This rotation of the tools ensures the drilling of a round hole. Another function of the rope socket is to provide, by its weight, part of the energy of the upward blows of the jars.

The components of the tool

67

string are usually joined together by tool joints of the box and pin type with standard American Petroleum Institute (API) designs and dimensions.

The bailer is simply a length of pipe with a check valve at the bottom. The valve may be either the flat-pattern or bell-and-tongue type called the dart valve. Fig. 5.16 shows a dart valve bailer being discharged by resting the tongue of the valve on a timber block.

Fig. 5.16 DISCHARGING DART VALVE BAILER.

The sand pump (Fig. 5.17) is a bailer fitted with a plunger which, when pulled upwards, creates a vacuum that opens the check valve and sucks the slurried cuttings into the bailer. Sand pumps are always made with flat-pattern check valves.

It is important that the drilling motion be kept in step with the fall of the string of tools for good operation. The driller must see to it that the engine speed has the same timing as the fall of the tools and the stretch of the cable. This is a skill that can only be provided by an experienced driller.

Drilling by the cable-tool percussion method in unconsolidated formations requires that the casing closely follows the drilling bit as the hole is deepened. This is necessary to prevent caving. The usual procedure is to dig a starting hole into which is placed the first section of casing. The casing is driven one to several feet into the formation, water added and the material within the casing drilled to a slurry and removed by bailing. The casing is then driven again and the material within it watered if necessary, drilled and removed by bailing. The procedure is repeated, adding lengths of casing until the desired depth is reached.

The pipe driving operation requires that the lower end of the first section of the casing be fitted with a protective casing shoe (Fig. 5.18). The top of the casing is fitted with a drive head which serves as an anvil. Drive clamps made of two heavy steel forgings and clamped to the upper wrench square of the drill stem are used as the hammer (Fig. 5.19). The string of tools, which provide the necessary weight for driving, is lifted and dropped repeatedly by the spudding action of the drilling machine, thus driving the casing into the ground. An alternative method of driving small diameter well casing uses a drive block assembly as previously shown in Fig. 5.4. The drive block is raised and dropped onto the drive head by means of manila rope wound on a cathead.

It is important that the first 40 to 60 feet of casing be driven vertically. Proper alignment of the string of tools centrally within the casing, when the tools are allowed to hang freely, is a necessary precaution. Periodic checks should be made with a plumb bob or carpenter's level used along the pipe at two positions approximately at right angles to each other to ensure that a straight and vertical hole is being drilled.

Cable-tool percussion drilling can be used successfully in all types of formations. It is, however, better suited than other methods to drilling in unconsolidated formations containing large rocks and boulders.

The main disadvantages of the cable-tool percussion method are its slow rate of drilling and the need to case the hole as drilling progresses. There are, however, a number of advantages that account for its widespread use. Reasonably accurate sampling of formation material can be readily achieved. Rough checks on the water quality and yield from each water-bearing stratum can readily be made as drilling proceeds. Much less water is needed for drilling than for the hydraulic rotary and jetting methods. This can be an important consideration in arid regions. Any encounter with water-bearing formations is readily noticed as the water seeps into the hole. The driller, therefore, need not be as skilled as his rotary counterpart in some respects.

INSTALLING WELL CASING

Some well drilling methods such as the cable-tool percussion method require that the casing closely follows the drill bit as drilling proceeds. In wells constructed by those methods, the casing is usually driven into position by any of the methods already described. This section deals with the setting of casing in an open borehole drilled by the hydraulic rotary, jetting, hydraulic percussion or sludger methods.

It is first necessary to ensure that the borehole is free from obstructions throughout its depth before attempting to set the casing. In the hydraulic rotary and jetting methods, the driller may ensure a clean hole by maintaining the fluid circulation with the bit near the bottom of the hole for a long enough period to bring all cuttings to the surface. At times, the driller may also drill the hole a little deeper than necessary so that any caving material fills the extra depth

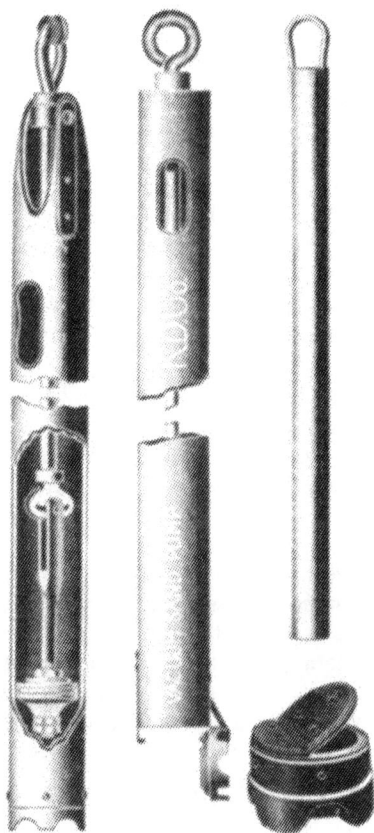

Fig. 5.17 SAND PUMP BAILER WITH FLAT VALVE BOTTOM.

69

Fig. 5.18 CASING DRIVE SHOE.

of the hole without affecting the setting of the casing at the desired depth.

In setting casing, it may be suspended from within a coupling at its top end by means of an adapter called a sub which is attached to a hoisting plug (Fig. 5.20), a casing elevator (Fig. 5.21) or a pipe clamp placed around the casing below the coupling. The first length of casing is lowered until the coupling, casing elevator or pipe clamp rests on the rotary table or other support placed on the ground around the casing. If lifting by means of a sub, the sub on the first length of casing is unscrewed and attached to the second length of casing. If lifting by elevators or pipe clamps, then the elevator bails or their equivalent are released from the casing in the hole and fixed to another elevator or pipe clamp on the second length of casing. This length of casing is then lifted into position and screwed into the coupling of the first length. The threads of the casing and coupling should be lightly coated with a thin oil. Joints should be tightly screwed together to prevent leakage. The elevator or other support for the casing is then removed and the string of casing lowered and supported at its uppermost coupling. The procedure is repeated for as many successive lengths of casing as are to be installed. Should caving be such as to prevent the lowering of the casing, the swivel may be attached to the casing with a sub and by circulating fluid through the casing wash it down. Alternatively, the casing may be driven.

Fig. 5.19 DRIVING CASING WITH DRIVE CLAMPS AS HAMMER AND DRIVE HEAD AS ANVIL.

GROUTING AND SEALING CASING

Grouting is the name given to the process by which a slurry or watery mixture of cement or clay is used to fill the annular space between the casing and the wall of the borehole to seal out contaminated waters from the surface and other strata above the desirable aquifer. Should the well be constructed with both an inner and outer permanent casing, then the space between the casings as well as that between the wall of the borehole and the outer casing should be grouted.

Puddled native clay of the type suitable for use as drilling fluid can

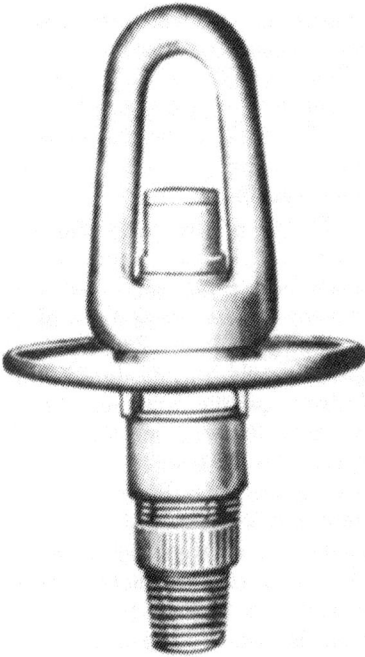

Fig. 5.20 HOISTING PLUG. (From Fig. 51 *Wells,* Department of the Army Technical Manual TM5-297, 1957)

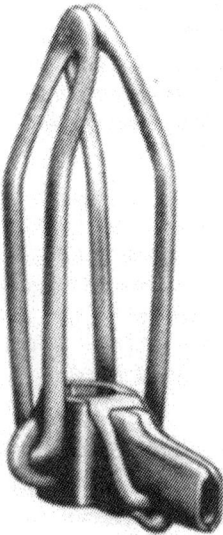

Fig. 5.21 CASING ELEVATOR.

be used for grouting and may be placed by pumping with the mud circulation pump normally used for drilling purposes. It should be used at depths below the first few feet from the surface where it would not be subject to drying and shrinkage. It should not be used at depths where water movement is likely to wash the clay particles away.

Cement grout is the type most commonly used and is the subject of the remainder of this section. It is made by mixing water and cement in the ratio of 5 to 6 gallons of water to a 94-lb sack of portland cement. This mixture is usually fluid enough to flow through grout pipes. Quantities of water much in excess of 6 gallons per sack of cement result in the settling out of the cement, which is undesirable. It is better to aim for the drier mixture based on the lower quantity of 5 gallons of water per sack of cement. A better flowing mixture may be obtained by adding 3 to 5 pounds of bentonite clay per sack of cement, in which case about 6.5 gallons of water per sack should be used. Where the space to be filled is large, sand may be added to the slurry to provide extra bulk. This, however, increases the difficulty of placing and handling. The water used in the mixture should be free of oil or other organic material such as plant leaves and bits of wood. Cement of either the regular or rapid-hardening type would be satisfactory. Use of the latter permits an earlier resumption of drilling operations

Mixing of the grout may be done in a concrete mixer, if available, and batches stored temporarily until enough is mixed for the job at hand. The quantities normally re-

71

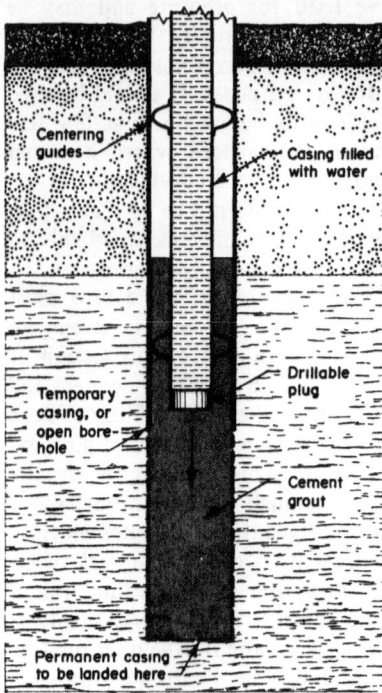

Fig. 5.22 A GRAVITY PLACEMENT METHOD OF CEMENT GROUTING WELL CASING. PLUGGED CASING LOWERED INTO CEMENT SLURRY FORCES SLURRY INTO ANNULAR SPACE.

quired for small wells can, however, be adequately mixed in a clean 50-gallon oil drum. To 20 gallons of water in the drum should be slowly sifted 4 sacks of cement while the water is being vigorously stirred with a paddle.

Placing of the grout should be carried out in one continuous operation before the initial set of the cement occurs. Regardless of the method of placing employed, the grout should be introduced at the bottom of the hole so that by working its way up the annular space fills it completely without leaving any gaps. Water or drilling mud should be pumped through the casing and up the annular space to clear it of any obstructions before placing the cement grout. To do this, the top of the casing must be suitably capped. If the borehole has been drilled much deeper than the depth to which the casing is being set, then the extra depth below the casing may be back-filled with a fine sand. There are several methods of placing grout, of which a few of the simpler ones are described below. Suitable pumps, air or water pressure may be used to force the grout into the annular space. However, grout may also be placed in shallow boreholes by gravity.

A *gravity placement method* is indicated in Fig. 5.22. A quantity of slurry in excess of that required to fill the annular space is introduced into the hole. The casing with its lower end plugged with easily drillable material (soft wood for example) and with centering guides is then lowered into the hole, forcing the slurry upwards through the annular space and out at the surface. The casing can be filled with water or weighted by other means to help it sink and displace the slurry. If temporary outer casing is used, it should be withdrawn while the grout is still fluid.

The *inside-tubing* method for grouting well casing is shown in Fig. 5.23. The grout is placed in the bottom of the hole through a grout pipe set inside the casing and is forced up the annular space either by gravity, or preferably by pumped pressure in order to complete the operation before the initial set of the cement occurs. Grouting must be continued until the slurry overflows

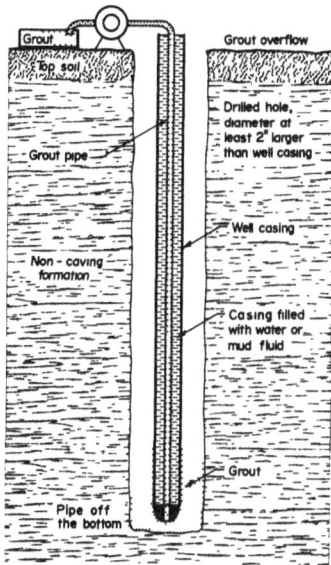

Fig. 5.23 INSIDE-TUBING METHOD OF CEMENT GROUTING WELL CASING.

the top of the borehole. A suitable packer or cement plug fitted with a ball valve is provided to the bottom end of casing to prevent leakage of the grout up the inside of the casing. This packer too must be made of easily drillable materials. The grout pipe should be ¾ inch or larger in diameter and the casing filled with water to prevent it from floating. The diameter of the drilled hole should be at least 2 inches larger than that of the well casing.

The *outside-tubing method* shown in Fig. 5.24 requires a borehole 4 to 6 inches larger in diameter than the well casing. The casing must be centered in the hole and allowed to rest on the bottom of it. The grout pipe, of similar size to that used in the inside-tubing method, is initially extended to the bottom of the annular space and should remain submerged in the slurry throughout the placing operations. This pipe may be gradually withdrawn as the slurry rises in the annular space. Should grouting operations be interrupted for any reason, the grout pipe should be withdrawn above the placed grout. Before lowering the pipe into the slurry again, grout should be used to displace any air and water in the pipe. The slurry is best placed by pumping, though it can be done by gravity flow. The casing may be plugged and weighted with water to prevent it from floating. The weight of the drilling tools may also be used to keep the casing in place.

After cement grout has been placed, no further work should be done on the well until the grout has hardened. The time required for hardening may be determined by placing a sample of the grout in an open can and submerging it in a bucket of water. When the sample has firmly hardened, work may proceed. Generally, a period of at least 72 hours should be allowed for cement grout to harden. If rapid-hardening cement is used, the time may be reduced to about 36 hours.

WELL ALIGNMENT

Alignment is being used here to include both the concepts of plumbness and straightness of a well. It is important to understand these concepts and how they differ. *Plumbness* refers to the variation with depth of the center line of the well from the vertical line drawn through the center of the well at the top of the casing. *Straightness*, however, merely considers whether the center line of the well is straight or otherwise. Thus, a well may be straight

73

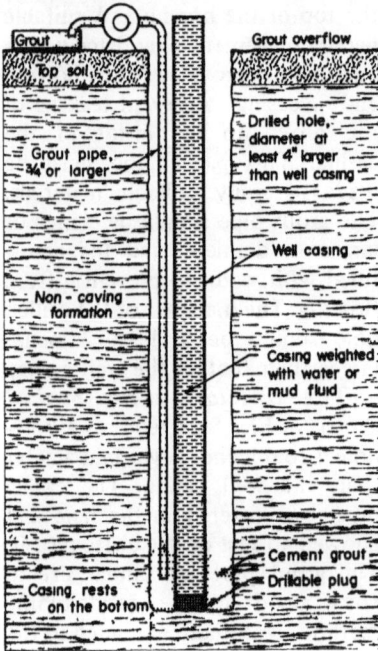

Fig. 5.24 OUTSIDE-TUBING METHOD OF CEMENT GROUTING WELL CASING.

but not plumb, since its alignment is displaced in some direction or other from the vertical.

Plumbness and straightness of a well are important considerations of well construction because they determine whether a vertical turbine or submersible pump of a given size can be installed in the well at a given depth. In this respect, straightness is the more important factor. While a vertical pump can be installed in a reasonably straight well that is not plumb, it cannot be installed in a well that is crooked beyond a certain limit. Plumbness must, however, be controlled within reasonable limits, since the deviation from the vertical can affect the operation and life of some pumps. Most well construction codes and drilling contracts specify limits for the alignment of large diameter, deep wells. Generally, these limits cannot be practicably applied to small diameter, shallow wells. These latter wells should merely be required to be sufficiently straight and plumb to permit the installation and operation of the pumping equipment.

Conditions Affecting Well Alignment

While it is desirable that a well be absolutely straight and plumb, this ideal is not usually achievable. Various conditions such as the character of the subsurface material being drilled, the trueness or straightness of the drill pipe and the well casing, and the pulldown force on the drill pipe in rotary drilling combine to cause variations from true straightness and plumbness. Varying hardness of materials being penetrated can deflect the bit from the vertical. So can boulders encountered in glacial drift formations. A straight hole cannot be drilled with crooked drill pipe. Too much force applied at the top end of the rotary drill stem will bend the slender column of drill pipe and cause a crooked hole. Weight, in the form of drill collars, placed at the lower end of the drill stem just above the bit, however, will help to overcome the tendency to drift away from the vertical. Even after the borehole is drilled, bent or crooked casing pipes and badly aligned threads on them can result in a well with appreciable variations from the vertical and straight lines.

74

Measurement of Well Alignment

Measurement of alignment is usually done in the cased borehole. When drilling has been by the rotary method these measurements should be made before the casing is grouted and sealed. For the cable-tool percussion and other methods in which the casing follows the bit as drilling progresses, periodic checks can be made on the plumbness and straightness during drilling. When a cable-tool hole has been started with the tools suspended directly over the center of the top of the casing, then any subsequent deviation of the cable from the center indicates a deviation of the hole from the vertical. The wearing of the corners of the cable-tool percussion bit on one side only also serves to indicate that a crooked hole is being drilled. These early indications help a driller to take steps to correct the fault. He may find it necessary to change the position of the drilling rig or backfill a portion of the hole and redrill it.

A plumb bob suspended by wire cable from the derrick of the drilling rig or from a tripod is usually used to measure both straightness and plumbness of a well. The plumb bob should be in the form of a cylinder 4 to 6 inches long with outside diameter about ¼ inch smaller than the inside diameter of the casing. It should be heavy enough to stretch the wire cable taut. A guide block is fixed to the derrick or tripod so that the center of its small sheave or pulley is 10 feet above the top of the casing and adjusted so that the plumb bob hangs exactly in the center of the casing. The wire cable should be accurately marked at 10-ft. intervals.

When the plumb bob is lowered to a particular 10-ft mark below the top of the casing the measured deviation of the wire line from the center of the top of the casing multiplied by a number that is one unit larger than that of the number of 10-ft sections of cable in the casing gives the deviation at the depth of the plumb bob. For example, if the deviation from the center at the top of the casing is 1/8 inch when the plumb bob is 30 feet below the top of the casing, then the deviation from the vertical at 30 feet depth in the casing is three plus one, or four, times 1/8 inch, that is 1/2 inch. Similarly, with the plumb bob 40 feet in the hole, the multiplier is five, and when 100 feet, the multiplier is eleven.

To determine the straightness, the deviation is measured at 10-ft intervals in the well. If the deviation from the vertical increases by the same amount for each succeeding 10-ft interval, then the well is straight as far as the last depth checked. The calculated deviation or drift from the vertical may be plotted against depth to give a graph of the position of the axis or center line of the well. Such a graph can be used to determine whether a pump of given length and diameter can be placed at a given depth in the well. This can also be checked on site by lowering into the well a "dummy" length of pipe of the same dimensions as the pump.

INSTALLATION OF WELL SCREENS

There are several methods of installing well screens, some of which are described below. The choice of method for a particular well may be influenced by the design of the well, the drilling method and the type of problems encountered in the drilling operation.

Pull-back Method

The pull-back method is by far the safest and simplest method used. While it is commonly used in wells drilled by the cable-tool percussion method, it is equally applicable in rotary drilled wells. The screen is lowered within the casing, which is then pulled back a sufficient distance to expose the screen. The screen must be the telescope type with outside diameter sized just sufficiently smaller than the inside diameter of the casing to permit the telescoping of the screen through the casing. The top of the screen is fitted with a lead packer which is swedged out to make a sand-tight seal between the top of the screen and the inside of the casing.

The basic operations in setting a well screen by the pull-back method are indicated in the series of illustrations in Fig. 5.25. The casing is first sunk to the depth at which the bottom of the screen is to be set. Any sand or other cuttings in the casing must be removed by bailing or washing. The screen is then assembled, suspended within the casing, following which the hook shown in Fig. 5.26 is caught in the bail handle at the bottom of the screen. The whole assembly is then lowered on the hoist line to the bottom of the hole. If the depth to water level in the hole is less than 30 feet, however, the assembled screen may simply be dropped in the casing. Having checked to

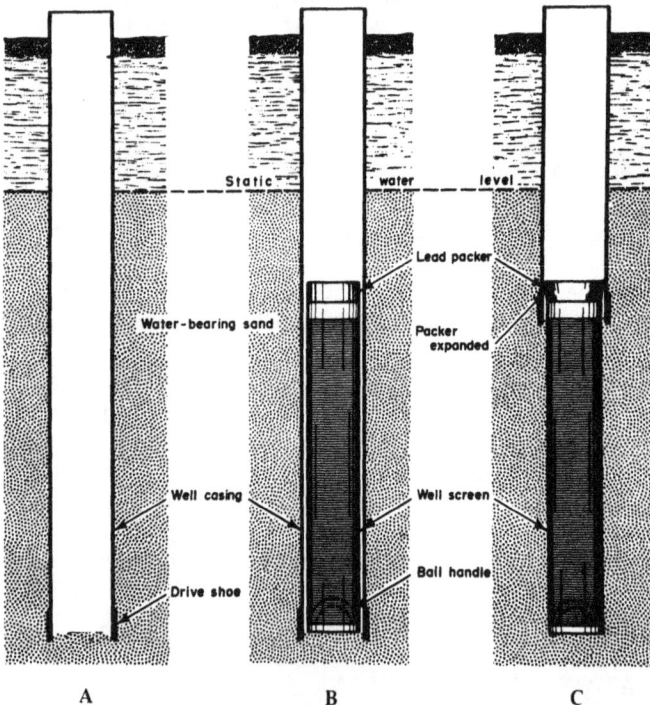

Fig. 5.25 PULL-BACK METHOD OF SETTING WELL SCREENS.
A. Casing is sunk to full depth of well.
B. Well screen is lowered inside casing.
C. Casing is pulled back to expose screen in water-bearing formation.

76

Fig. 5.26 LOWERING HOOK.

Fig. 5.27 B U M P I N G BLOCK BEING USED TO PULL WELL CAS-ING. (From Bergerson-Caswell, Inc , Minneapolis, Minnesota.)

ascertain the exact position of screen, the hook is released and withdrawn. A string of small pipe is then run into and allowed to rest on the bottom of the screen to hold it in place while the casing is being pulled back to expose the screen. If the casing has been driven by the cable-tool percussion method, then it may be pulled by jarring with the drilling tools or with a bumping block, the latter of which is shown in Fig. 5.27. It may even be possible in some instances to pull the casing with the casing line on the drilling machine. Mechanical or hydraulic jacks (Fig. 5.28) may also be used in combination with a pulling ring or spider with wedges or slips. The casing should be pulled back far enough to leave its bottom end 6 inches to 1 foot below the lead packer. The pipe holding the screen in place is removed and a swedge block (Fig. 5.29) used to expand the lead packer and create a sand-tight seal against the inside of the casing. To do this, two or three lengths of small diameter pipe are screwed to the sliding bar which passes through the swedge block. The assembly is lowered into the well until the swedge block rests on the lead packer. The weight provided by the pipe attached to the sliding bar is then lifted 6 to 8 inches and dropped several times. The swedge block itself should not be lifted off the lead packer. It should be simply forced down into the packer by the repeated blows of the weighted sliding bar.

Open Hole Method

The open hole method illustrated in Fig. 5.30 involves the setting of the screen in an open hole drilled below the previously installed casing. The method is applicable to rotary drilled wells.

Fig. 5.28 PULLING CASING BY HYDRAULIC JACKS IN COMBINATION WITH CAS· ING RING AND SWEDGE SLIPS.

Fig. 5.29 SWEDGE BLOCK.

The borehole is first drilled to the depth at which the casing is to be set permanently. The casing is run into the hole and grouted as required. Using a bit just large enough to go through the casing, the hole is drilled through the water-bearing formation below the casing. A suitable drilling mud must be used to seal off all flow from the formation into the hole, prevent it from caving, and transport the cuttings out of it. Fluid circulation must be maintained long enough after the desired depth is reached to ensure that all cuttings have been lifted from the hole. The drill stem may then be withdrawn and a telescope-size screen lowered into the hole by any convenient means. The depth of the hole should first have been checked to ensure that, with the screen resting on the bottom of the hole, the lead packer remains inside the lower end of the casing. Gravel may be used to back-fill a hole that has been drilled too deep. A screen with a closed bail bottom can be set by this method provided the precautions have been taken to obtain a non-caving hole free from cuttings and a suitable drilling mud that does not allow cuttings to settle out before the screen is lowered into the well. If difficulties are experienced in

78

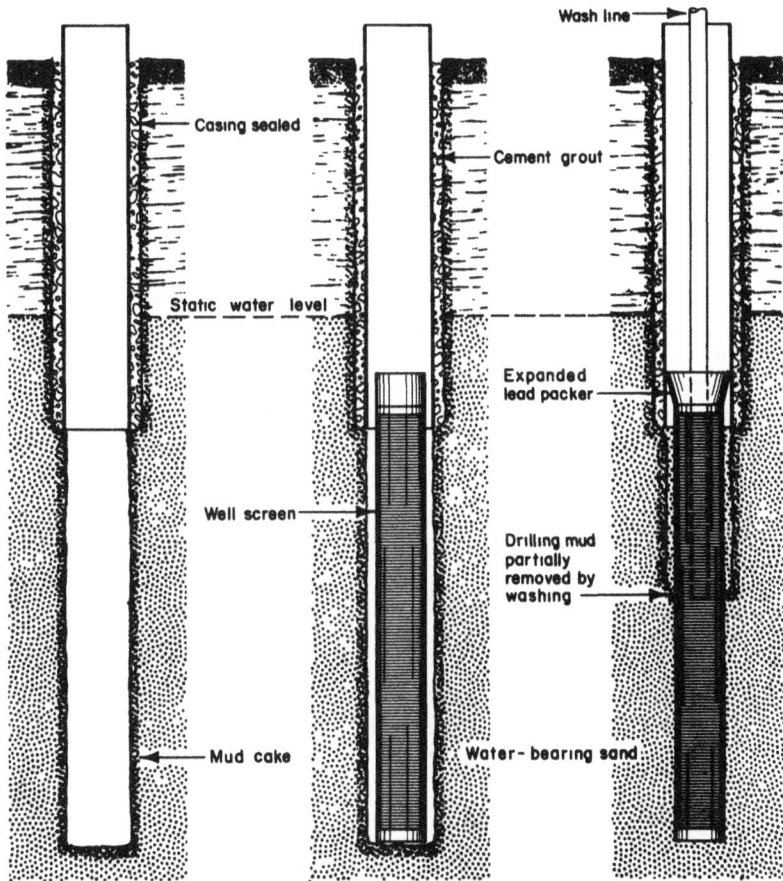

Fig. 5.30 SETTING WELL SCREEN IN OPEN HOLE DRILLED BELOW THE WELL CASING.

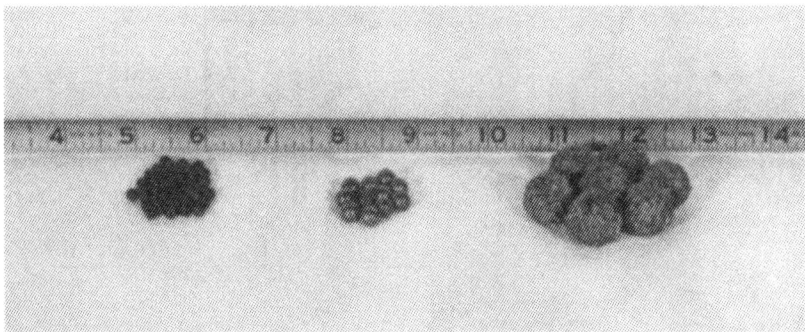

Fig. 5.31 LEAD SHOT AND LEAD WOOL FOR PLUGGING OPEN BOTTOM END OF WELL SCREEN.

maintaining such a "clean" hole, a short extension pipe may be attached to the bottom of an open-ended screen to permit washing it down with drilling fluid. The bottom of the extension pipe is then plugged with lead shot, lead wool (Fig. 5.31) or cement grout and the lead packer expanded after circulating water to wash some of the drilling mud out of the hole. Lead wool or cement grout should be tamped for compaction. If lead shot is used, it is simply poured in sufficient quantity to form a 4- to 8-inch thick layer inside the extension pipe.

Wash-down Method

The wash-down method of installation (Fig. 5.32) uses a high velocity jet of light-weight drilling mud or water issuing from a special wash-down bottom fitted to the end of the screen to loosen the sand and create a hole in which the screen is lowered.

The wash-down bottom is a self-closing ball valve. A string of wash pipe is connected to it and used to lower the entire screen assembly through the casing which has been previously cemented. As the screen is washed into position, the loosened sand rises around the screen and up through the casing

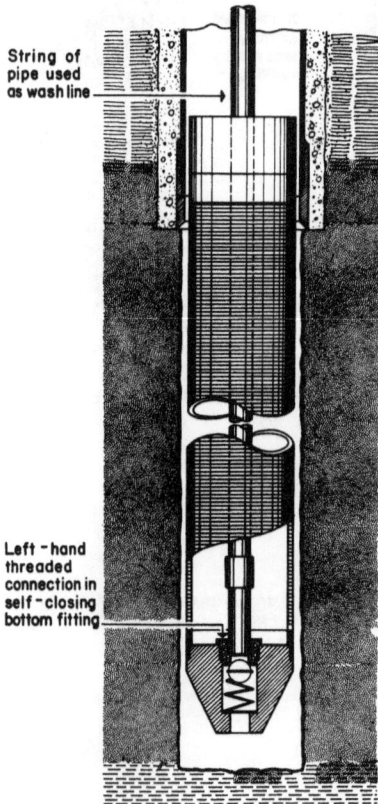

Fig. 5.32 WASH-DOWN METHOD OF SETTING WELL SCREEN.

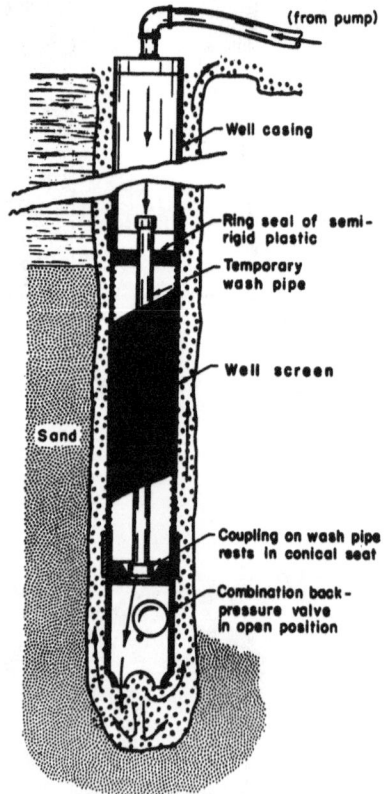

Fig. 5.33 JETTING WELL SCREEN INTO POSITION.

to the surface with the return flow. Sand particles which inevitably accumulate in the well screen must be washed out of it once the screen is in final position. Water should later be circulated at a reduced rate to remove any wall cake formed in the hole during the jetting operation. This causes the formation to cave around the screen and grip it firmly enough for the wash line to be disconnected.

It is common practice in jetted and rotary drilled small wells to set a combined string of casing and screen, permanently attached, in one operation. A jetting method for setting such a combined string is illustrated in Fig. 5.33. The scheme employs the use of a temporary wash pipe assembled inside the well screen before attaching the screen to the bottom length of casing. A coupling attached to the lower end of the wash pipe rests in the conical seat in the wash-down bottom. A close-fitting ring seal made of semi-rigid plastic material or wood faced with rubber is fitted over the top end of the wash pipe and kept in position by the coupling above it. The seal prevents any return flow of the jetting water in the space between the wash pipe and the screen. All the return flow from the washing or jetting operation, therefore, takes place outside of the screen and casing. A little leakage of the jetting water takes place around the bottom of the wash-pipe and out through the screen, thus preventing the entry of fine sand into the screen. Maintaining this small outward flow through the screen is important, since it reduces the possibility of sand-locking the wash pipe in the screen.

With the casing and screen assembly washed into final position, fluid circulation is stopped. The plastic ball then floats up into the seat, thus effectively closing the valve opening in the washdown bottom. A tapered tap, overshot or some other suitable fishing tool (see later section of this chapter on fishing tools) is then used to fish the wash pipe and ring seal out of the screen. It may also be possible to recover the wash pipe assembly by tapping the coupling with pipe carrying regular pipe threads instead of a tapered tap. The well is then ready for development.

Satisfactory penetration by this method requires continuous circulation when water is used as the jetting fluid. This may limit the use of the method to the penetration of only as much screen and casing as it is physically possible to assemble as a single string in an upright position with the available drilling equipment. Subsequent additions of casing will require interruptions of the circulation that can lead to the collapse of the drill hole (particularly in water-bearing sands and gravels) around the combined string of screen and casing thus preventing further penetration. This problem may be avoided by the use of a suitable drilling mud. The method is very often used for washing screens into position below previously drilled boreholes. If the borehole has already been drilled into the aquifer to the full depth of the well, then the wash-down bottom may be used on the screen without the wash pipe.

Well Points

Well points can be and are often installed in drilled wells by some of the methods just described in this section. The pull-back and open hole methods would be particularly applicable. Where, because of excessive friction on the casing or a heaving sand formation, the pull-back method is impracticable, a well point may be driven into the formation below the casing by either of the methods shown in Fig. 5.34 or Fig. 5.35. In the method of Fig. 5.35 the driving force is transmitted through the driving pipe directly onto the solid point of the screen. This method is preferable, therefore, when driving relatively long well points. In both cases the hole is kept full of water while the screen is being set in heaving sand formations.

Artificially Gravel-Packed Wells

The methods of screen installation so far described apply primarily to wells to be completed by natural development of the sand formation. One of these, the pull-back method, can, with little modification, be used in artificially gravel-packed wells.

An artificially gravel-packed well has an envelope of specially graded sand

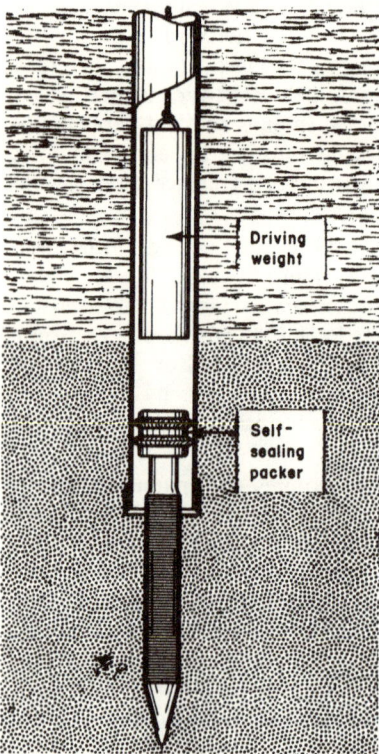

Fig. 5.34 DRIVING WELL POINT WITH SELF-SEALING PACKER INTO WATER-BEARING FORMATION.

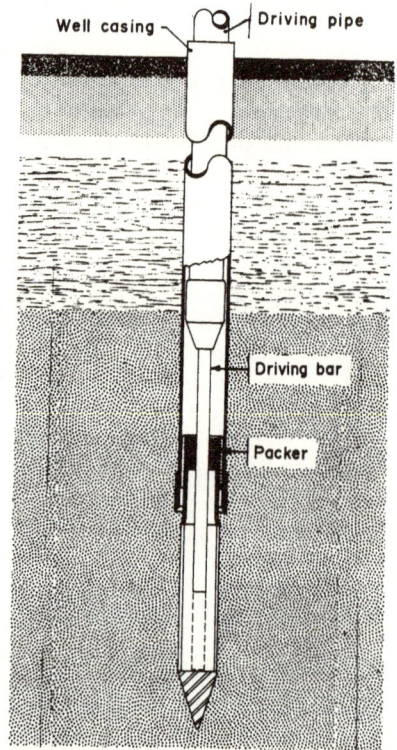

Fig. 5.35 DRIVING BAR USED TO DELIVER DRIVING FORCE DIRECTLY ON SOLID BOTTOM OF WELL POINTS 5 FT OR MORE IN LENGTH.

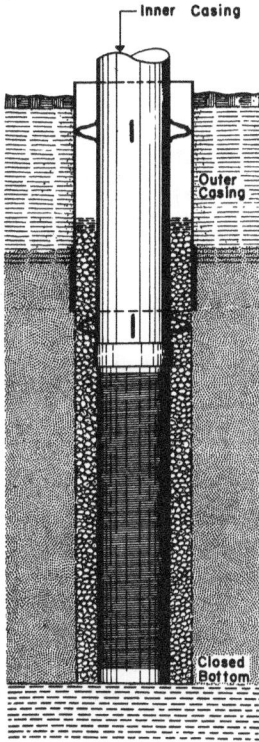

Inner Casing

Outer Casing

Closed Bottom

Fig. 5.36 DOUBLE-CASING METHOD
OF ARTIFICIALLY GRAVEL
PACKING A WELL. GRAVEL
IS ADDED AS THE OUTER
CASING IS PULLED BACK
FROM THE FULL DEPTH
OF THE WELL:

or gravel placed around the well .screen in a predetermined thickness. This envelope takes the place of the hydraulically graded zone of highly permeable material produced by conventional development precedures. Conditions that require the use of artificial gravel packing have been described in the previous chapter.

The modified pull-back method known as the double-casing method involves centering a string of casing and screen of equal diameter within an outer casing of a size corresponding to the outside diameter of the gravel pack (Fig. 5.36). This outer casing is first set to the full depth of the well. The inner casing and screen should be suspended from the surface until the placement of the gravel pack is completed. The selected gravel is put in place in the annular space around the screen in batches of a few feet, following each of which the outer casing is pulled back an appropriate distance and the procedure repeated until the level of the gravel is well above the top of the screen. The well may then be developed to remove any fine sand from the gravel and any mud cake that may have formed on the surface between the gravel and the natural formation. The method can be used in both cable-tool percussion and rotary drilled wells.

Care must be taken in placing the gravel to avoid separation of the coarse and fine particles of the graded mixture. Failure to do so could result in a well that continually produces fine sand even though properly graded material has been used in the gravel pack. This tendency towards separation of particles of different sizes can be overcome by dropping the material in small batches or slugs through the confined space of a small diameter conductor pipe or tremie (Fig. 5.37). Under these confined conditions there is less tendency for the grains to fall individually. Water is added with the gravel to avoid bridging in the tremie. The tremie, usually about 2 inches in diameter, is raised as the level of material builds up around the well screen. Water circulated in a reverse direction to that of normal rotary drilling — that is down the annular space between the casings, through the gravel and screen and up through the inner casing to the pump suction — helps prevent bridging in the annular

Fig. 5.37 PLACING GRAVEL−PACK MATERIAL THROUGH PIPE USED AS TREMIE.

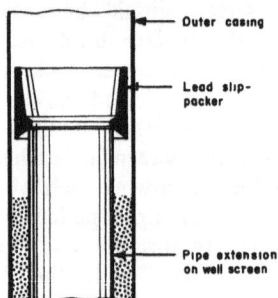

- Outer casing
- Lead slip-packer
- Pipe extension on well screen

Fig. 5.38 LEAD SLIP-PACKER IN PO-SITION ON EXTENSION PIPE BEFORE EXPANSION TO SEAL THE ANNULAR SPACE.

space as the gravel is being deposited.

Some settlement of the gravel will occur during the development process. More gravel must, therefore, be added as is necessary to keep the top level of the gravel several feet above that of the screen. The entire length of the inner casing need not be left permanently in the well if the outer one is intended to be permanent. Towards this end, a convenient joint in the inner casing can be loosely made up while assembling the string. After development of the well the upper portion of casing is then unscrewed at this joint and withdrawn, leaving enough pipe (at least one length) attached to the screen to provide an overlap of a few feet within the outer casing.

Another technique would be to set the inner casing to the full depth of the well and telescope the screen and an appropriate length of extension pipe attached to the top of the screen into that casing. The entire string of inner casing may then be removed as the gravel is placed, leaving the extension pipe overlapping inside the outer casing. Centering guides must be provided on the temporary inner casing.

Cement grout, lead shot or pellets of lead wool can be used to seal the annular space immediately above the top of the gravel. A mechanical type of seal known as a lead slip-packer (Fig. 5.38) is also often used. The packer, a lead ring of similar shape to a casing shoe, sits on top of the extension pipe and is of the proper diameter and wall thickness to form an effective seal when expanded by a swedge block against the outer casing.

Recovering Well Screens

It may sometimes be necessary to recover an encrusted screen for cleaning and return to the well, a badly corroded one for replacement or a good one from an abandoned well for reuse elsewhere. Considerable force may have to be applied to the screen to overcome the grip of the water-bearing sand around it. The sand-joint method provides one of the best ways of transmitting this force to the screen, dislodging and recovering it without damaging it. The

84

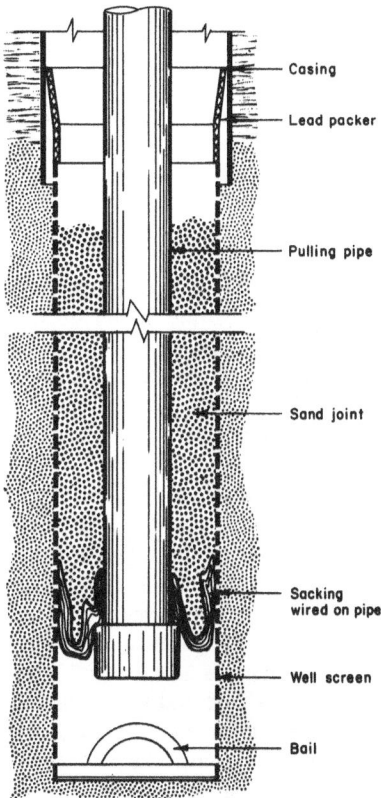

Casing

Lead packer

Pulling pipe

Sand joint

Sacking
wired on pipe

Well screen

Bail

Fig. 5.39 ELEMENTS OF SAND-JOINT
METHOD USED FOR PUL-
LING WELL SCREENS.

method, however, cannot be used in screens smaller than 4 inches in diameter.

The *sand-joint* method uses sand carefully placed in the annular space between a pulling pipe and the inside of the well screen to form a sand lock or sand joint which serves as the structural connection between the pulling pipe and the screen (Fig. 5.39). The necessary upward force may then be applied to the pulling pipe by means of jacks working against pipe clamps or a pulling ring with slips as shown in Fig. 5.28.

The size of the pulling pipe varies with the diameter of the screen and the force which may be required. As a general rule, however, the size of pipe is chosen at one-half the nominal inside diameter of the screen. For example, a 4-inch screen with nominal inside diameter of 3 inches would require 1½-inch pipe. Extra heavy pipe should be used. The pipe couplings and threads should be of the highest quality in order to withstand the pulling force. The sand should be clean, sharp and uniform material of medium to moderately fine size.

The first step in the preparation of the sand joint is the tying of 2-inch strips of sacking to the lower end of the pulling pipe immediately above a coupling or ring welded to the pipe (Fig. 5.40). The sacking forms a socket to retain the sand fill around the pulling pipe. The pipe and sacking with both ends tied to the pipe are then lowered into the casing until only the upper ends of the strips remain above the top of the casing. The string which holds the upper ends of the sacking to the pipe is then cut and the strips of sacking arranged evenly around the top of the casing as shown in Fig. 5.41.

Next the pulling pipe is lowered to a point near the bottom of the screen, care being taken to keep it as well centered as possible. The sand is then poured slowly into the annular space between the pulling pipe and the casing. An even distribution of the sand around the circumference of the pipe is desirable. The pulling pipe should be moved gently backward and forward at the top while pouring the sand to avoid bridging above couplings. A small stream of water playing onto the sand would also help in preventing bridging. Enough

85

sand should be used to fill at least two-thirds but not the entire length of the screen. The level of the sand in the screen can be checked with a string of small diameter pipe used as a sounding rod.

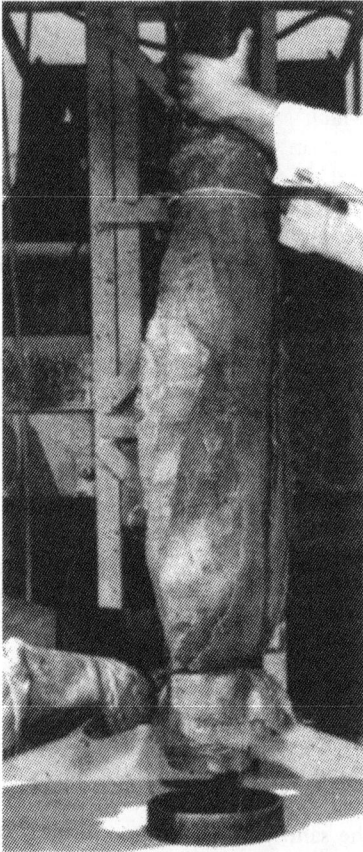

Fig. 5.40 STRIPS OF SACKING BEING TIED TO LOWER END OF THE PULLING PIPE USED IN THE SAND-JOINT METHOD.

The proper quantity of sand having been placed, the pulling pipe is then gradually lifted to compact the sand and develop a firm grip on the inside surface of the screen. Additional tension is applied until the screen begins to move. The screen may then be pulled steadily without difficulty until it is out of the well. The sand joint can be broken at the surface by washing out the sand with a stream of water.

Pre-treatment of the screen with hydrochloric or muriatic acid serves to loosen encrusting materials and thus reduce the force required to obtain initial movement of the screen. For this purpose the screen is filled with a mixture of equal amounts of acid and water which is left to stand for several hours, overnight if convenient. The acid is then pumped or bailed out before starting the pulling operations.

FISHING OPERATIONS

A *fish* is the name used collectively to describe a well drilling tool, length of casing or other similar equipment or material accidentally deposited or stuck in boreholes and wells and which it is desirable to recover. Several reasons may contribute to the desirability for recovering a fish. For instance, the nature and position of the fish may be such as to prevent further work on the borehole towards the completion of a well. The fish may be a tool, a piece of equipment or material which is vital to the drilling operations and, in addition, costly and not easily replaceable. Fishing operations involve a considerable element of trial and error, because the fish is out of sight at some depth in a hole. They can, therefore, be very time and cost consuming with no guarantee of success. Consequently, very careful consideration should be given to the possible cost of a fishing operation in terms of time and money, comparing this with the losses and time

86

saving that abandonment of the borehole or well would entail. Only after such careful consideration should fishing operations be undertaken. For small diameter, relatively shallow wells it would often be found economical and otherwise beneficial to drill a new well rather than attempt fishing operations in one under construction. This is particularly true prior to the placing and cementing of the permanent casing. It should also be borne in mind that fishing operations require a great deal of skill, much more so than drilling operations and the driller may be inexperienced in such work.

Preventive Measures

As is the case with all other forms of accidents, prevention is always better than cure. Towards this end, the necessity to exercise the greatest care and attention at all times and throughout all stages of drilling operations cannot be over-stressed. While the utmost care and attention will not completely eliminate the need for fishing, it will considerably reduce the number and frequency of fishing operations.

Among the precautions that should be undertaken is the proper care and use of drilling tools and equipment. This includes the proper cleaning and breaking-in of new tool joints, the proper cleaning and setting of joints at all times, the correct dressing and hardening of bits, the regular maintenance and inspection of all wire rope, the regular inspection of all components of the drilling string for the development of fatigue cracks and the discarding of worn out tools. Above all, care must be taken never to overload equipment nor use tools for purposes other than those for which they have been designed. The manufacturer's limitations set on the use of equipment and tools should not be exceeded.

The care of wire rope should be given special consideration. Many manufacturer's catalogs contain detailed instructions. Among the most important of these is the need for regular lubrication with a good grade of lubricant, free from acid or

Fig. 5.41 UPPER END OF SACKING STRIPS ARRANGED EVEN-LY AROUND TOP OF WELL CASING AS THE PULLING PIPE IS LOWERED INTO THE WELL.

87

alkali and which will penetrate and adhere to the rope. The use of crude oil or other material likely to be injurious to steel or cause deterioration or brittleness of the wires must be avoided. Failure to properly lubricate wire rope results in the wires becoming brittle, corroded, subject to excessive friction wear and ultimately the sudden fracturing of the rope. The rope should be tightly and evenly wound on winding drums and should not be allowed to stand in mud, dirt or other such medium which is harmful to steel. Only proper fastening clamps that do not kink, flatten or crush the rope should be used. The fracturing of loaded wire rope, it should be remembered, can cause serious injury to workmen as well as create fishing problems.

Unscrewed tool joints are the causes of many fishing operations. These can be avoided by the proper mating of the box and pin components of the joints. Both the pin shoulder and the box face should be thoroughly cleaned and free of imperfections that prevent a full and even contact. The threads and shoulders of the component parts should be thinly coated with a light machine oil before making up the joint. Joints should be firmly made up though not with excessive pressure as this can result in broken boxes and pins.

Tools, carelessly left on the rotary table or at some such point, may be accidentally tipped into a borehole. One half of a pipe clamp entering a well in this manner has been known to become wedged in the well screen just above a joint in the wash pipe being used in the development process and result not only in the abandonment of the well but also the loss of several hundred feet of drill stem with it. All tools should be removed immediately after use to a convenient point of storage at a safe distance from the borehole or well.

Certain conditions such as slanting or caving formations, crooked holes and the presence of boulders often contribute to drilling troubles that may result in fishing operations. The utmost care must be exercised by drillers operating under these conditions.

Preparations for Fishing

The nature of all operations (construction and maintenance) on wells is such that accidents do occur even under the supervision of the most capable and careful drillers. Therefore, the driller in anticipation of the inevitable fishing job should record or have access to the exact dimensions of everything used in or around the well. This facilitates the selection and design of a suitable fishing tool when necessary. All tools brought to the site should be accurately measured and the measurements properly recorded. Some of the important measurements are: the outside diameter and length of the rope socket; the diameter, length and stroke of the drill jars; the diameter and length of the drill stem; the size of tool joints and the outside diameter and length of the pin and box collars; the body size and length of bits; the length of pin collars on the bits. A careful record of the depth of the hole and the overall length of the drilling string is also essential for successful fishing operations.

The drill hole must of necessity be larger than any tools placed in it. As a result, tools lost in a hole do not often remain in the vertical or upright position but become wedged in sloping positions across the hole. In addition, material from a caving formation may fall onto and cover the tool. No amount of measurement at the surface could tell the driller exactly what position the lost tool has assumed in the hole or, in some cases, whether the top portion of it is free from obstruction. It is, therefore, considered good practice to use what is known as an *impression block* to obtain an impression of the top of the tool before attempting any fishing operations. This is particularly necessary in rotary drilled, uncased holes. Impression blocks are of many forms and designs, one of which is shown in Fig. 5.42. A short block of wood (preferably soft wood) turned on a lathe to a diameter about one inch less than that of the drilled hole and with the upper portion shaped in the form of a pin, is driven to fit tightly into a drill pipe box collar. For added security, the wooden block should be wired or pinned securely to the collar. The wooden block may alternatively be bolted to the dart of a dart valve bailer. A quantity of small headed nails is driven into the bottom of the circular block, leaving an extension of about ½ inch. Sheet metal is temporarily nailed around the block with a protrusion of a few inches over the lower end of the block. Warm paraffin wax, yellow soap or other plastic material is poured to fill this protrusion and then left to cool and solidify. The nail heads help to hold the plastic material onto the block. After the sheet metal is removed and the lower end of the plastic material carefully smoothed, the impression block is ready for use. The block should be lowered carefully and slowly into the hole until the object is reached. It is then raised to the surface where the impression made in the wax or soap can be examined. By careful

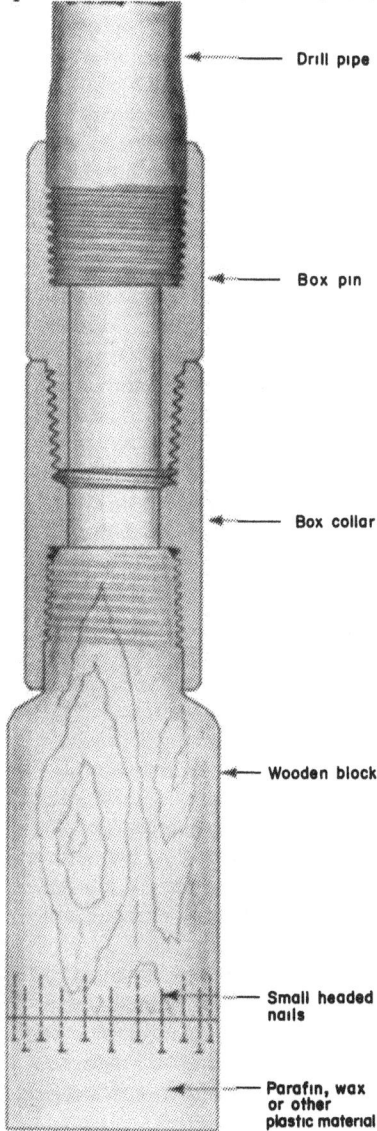

Fig. 5.42 IMPRESSION BLOCK.

Drill pipe

Box pin

Box collar

Wooden block

Small headed nails

Parafin, wax or other plastic material

interpretation of the impression, a driller can determine the position of the fish and the best means of retrieving it.

Common Fishing Jobs and Tools

It is often said, with considerable justification, that no two fishing jobs are alike. While fishing jobs may be classified into various types, individual jobs within these types are usually quite different. Fishing jobs, as a result, test the skill and ingenuity of the driller to the fullest extent. The driller relies on a number of basic principles in his attack on fishing problems. A great variety of special tools have been devised to assist him in this work. Many of these tools are used very infrequently and it is not uncommon to find a tool made for a particular job and never used again. Only large-scale drilling operators can afford to have more than a limited stock of fishing tools. Whenever possible, small operators usually rent tools as they are needed from suppliers. It would be impractical to attempt a discussion of all types of fishing jobs and the tools used on them. Instead, the discussion that follows centers on some of the more common types of fishing jobs and tools.

(1) *Parted drill pipe:* One of the most frequent fishing jobs in rotary drilling is that for the recovery of drill pipe twisted off in the hole. The break may either be due to shearing of the pipe or failure of a threaded joint.

An impression block should first be used to determine the exact depth and position of the top of the pipe, whether there has been any caving of the upper formation material onto the top of the pipe or whether the pipe has become embedded into the wall of the hole. If the top of the pipe is unobstructed, then either the *tapered fishing tap* or *die overshot* could be effective if used before the cuttings in the hole settle and "freeze" the drill pipe. The *circulating-slip overshot,* which permits the circulation of drilling fluid, would be the best tool to use after the pipe has been frozen by the settling of cuttings around it. These tools are all illustrated in Fig. 5.43.

The *tapered fishing tap,* made of heat-treated steel, tapers approximately 1 inch per foot from a diameter somewhat smaller than the inside diameter of the coupling to a diameter equal to the outside diameter of the drill stem. The tapered portion is threaded and fluted the full length of the taper to permit the escape of chips cut by the tap. The tap is lowered slowly on the drill stem until it engages the lost pipe, the circulation being maintained at a low rate through the hole in the tap during this period. Having engaged the lost pipe, the circulation is stopped and the tap turned slowly by the rotary mechanism or by hand until the tap is threaded into the pipe. An attempt should then be made to re-establish the circulation through the entire drill string before pulling the lost pipe.

The *die overshot* is a long-tapered die of heat-treated steel designed to fit over the top end of the lost drill pipe and cut its own thread as it is rotated. It is fluted to permit the escape of metal cuttings. Circulation cannot be completed to the bottom of the hole through the lost pipe since the flutes also allow the fluid to escape. The upper end of the tool has a box thread designed to fit the drill pipe.

The *circulating-slip overshot* is a tubular tool approximately 3 feet long with inside diameter slightly larger than the outside diameter of the drill pipe.

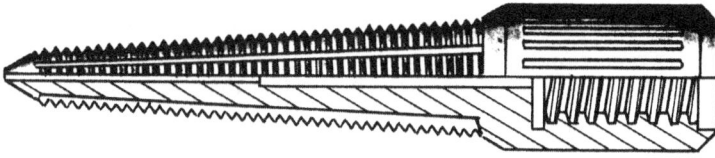

Tapered Tap

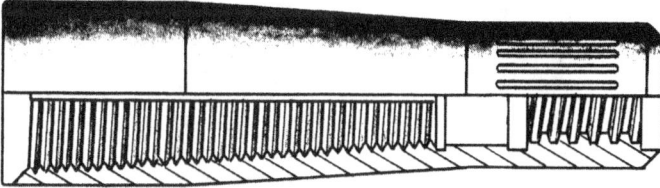

Die Overshot

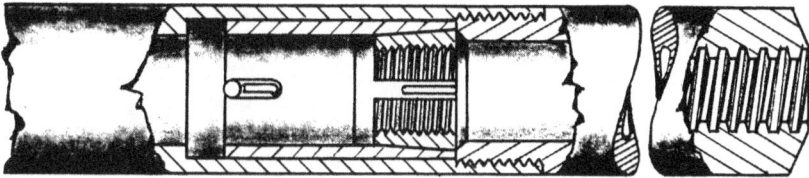

Circulating Slip Overshot

Fig. 5.43 TAPERED TAP AND OVERSHOTS. (Adapted from Fig. 43, *Wells*, Department of the Army Technical Manual TM5-297, 1957.)

The belled-out lower portion of the tool helps to centralize and guide the top of the lost drill pipe into the slip shown fitted in the tapered sleeve. The slot cut through one side of the slip enables it to expand as the tool is lowered over the drill pipe. As the tool is raised the slip is pulled down into the tapered sleeve, thus tightening the slip against the pipe. Circulation of fluid can then be established through the pipe, freeing it for recovery.

A *wall hook* shown in Fig. 5.44 can be used to set the lost drill pipe erect in the hole in preparation for the tap or overshot tools. The wall hook is a simple tool that can be made from a suitable size of steel casing cut to shape with a cutting torch. A reducing sub must then be used to connect the top end of the tool to the drill stem. To operate the wall hook, it is lowered until it engages the pipe, then slowly rotated until the pipe is fully within the hook. The hook is then raised slowly to set the pipe in an upright position, later disengaging itself from the pipe.

It is also possible to pin a tapered fishing tap into the upper portion of a wall hook made from steel casing. With such a combined tool, the hook may be used to realign the lost drill pipe and then, while being lowered, guide the tap into the drill pipe to complete both operations in one run of tools into the hole. This method is particularly desirable when the drill pipe tends to fall over against the wall of a much larger hole rather than remain erect.

Fig. 5.44 WALL HOOK.

(2) *Broken wire line:* When the drilling line or sand line of a cable-tool drilling rig breaks, leaving the drilling tools or bailer in the hole with a substantial amount of wire line on top of the tools, the *wire line center spear* (Fig. 5.45) is the recommended fishing tool. This tool consists of a single prong with a number of upturned spikes projecting from it. The spikes have sharp inside corners that permit the spear to catch even a single strand of wire. If the lost tools are stuck in the hole and cannot be pulled, the sharp spikes will shear the wire line.

The shoulder of the spear should be about the same size as the borehole in order to prevent the broken wire line from getting past the spear as it is lowered and causing it to become stuck in the hole. For the range of borehole sizes being considered, center spears are made for specific sizes of hole.

The spear is used with a set of fishing jars, short sinker and wire line socket above it. It should be carefully eased down the hole to the point where it is expected to engage the broken cable. It is then pulled to see if it has a hitch. In the absence of a hitch it is lowered below the first point and again tested for a hitch. This procedure is repeated until a hitch is secured.

If the string of lost tools is free, lift it 10 to 15 feet off the bottom of the hole and test the hold on the wire line by allowing the brake to give a short, quick slip. If the hold is insecure the tools will fall with no resulting damage, while a later fall through some greater distance could be very damaging.

Fig. 5.45 CENTER SPEAR.

If the hold is secure, continue lifting the tools out of the hole until the broken wires appear. Stop lifting and tie the wires together and then to the prongs of the grab to prevent the loose ends from unfolding and causing the hold to break. The tie itself does not carry the load but holds the broken lines in position. Continue lifting until the lost tools are recovered.

If the string of lost tools is not free, then sufficient line should be let out to bring the jars into use. Jarring should be continued until the lost tools come loose or the broken cable parts.

(3) *Fishing for the neck of a rope socket, other cylindrical object or the pin of a tool:* The *combination socket* (Fig. 5.46) is one of several tools used to catch the neck of a wire-line socket after broken line has been cleared away, or the pin of a bit or drill stem that has become unscrewed in the hole. The tool can also be used to fish for any cylindrical object such as a drill stem or tubing standing upright in the hole, providing the bore of the socket is at least 1/8 inch larger than the diameter of the fish. The fishing string should consist of a rope socket, stem, long-stroke fishing jars and combination socket.

Combination sockets are provided with two sets of slips, one set of which is used to engage the threads of the pin on a bit, stem or other tool and the second set to take hold of the neck of the rope socket. The proper set of slips must be selected for the particular fishing job in accordance with knowledge of the exact size of the fish. It is also good practice to determine if the socket can go over the fish by first running the socket with its inner parts removed. The re-loaded combination socket is

93

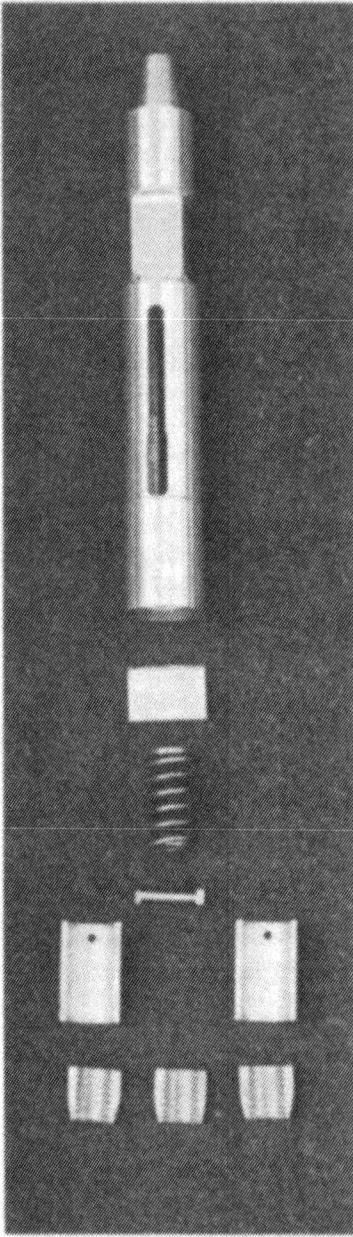

Fig. 5.46 COMBINATION SOCKET
DISMANTLED TO SHOW
SLIPS AND OTHER COMPO-
NENT PARTS. (From Acme
Fishing Tool Company,
Parkersburg, West Virginia)

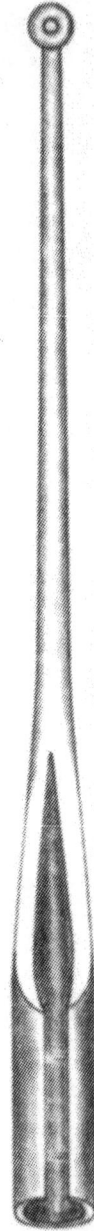

Fig. 5.47 JAR BUMPER

then slowly lowered on the fishing string with the fishing jars adjusted for shortest stroke. Upon contact with the fish, a light downward jar is used to secure a hitch. Tension is then taken on the line and the fishing job completed if the tools are not stuck.

If the tools are stuck, then a slow spudding action should first be tried to release them. Should this fail, then sufficient line is let out to bring the jars into use. Short and rapid jarring should cause the freeing of the tools and is preferable to hard long-stroke jarring even though several hours of work may be necessary. Long-stroke jarring could result in breaking of the hitch on the lost tools or in broken fishing tools. Alternate up-jarring and down-jarring would release the hitch on the lost tools, should it become obvious that they cannot be freed and recovered.

After successful completion of a fishing job, the hitch is broken by removing the wooden block above the spring in the combination socket and so relieving the pressure on the spring and slips.

(4) *Releasing locked jars:* Jars sometimes become stuck or tools above the jars wedged in the hole by a piece of rock or other material. A *jar bumper* (Fig. 5.47) is the tool normally used under such circumstances. The following procedure should be followed. A strain is first taken on the drilling cable. The jar bumper is then lowered on the sand line, using the drilling cable as a guide, until the bumper reaches the string of tools. The bumper is then raised 10 or 12 feet and dropped, repeating this as often as necessary to loosen the jars or string of tools. A few blows are usually sufficient for this purpose. Too many blows might batter the neck of the rope socket and should be avoided. Should the bumper fail to release the tools, cut the cable and use a combination socket.

CHAPTER 6

WELL COMPLETION

Well completion is the term used to describe the two basic processes which are undertaken after a well has been constructed in order to ensure a good yield of water that is clear and relatively free of suspended matter and disease-producing organisms. These processes are called *well development* and *well disinfection.*

WELL DEVELOPMENT

The object of well development is the removal of silt, fine sand and other such materials from a zone immediately around the well screen, thereby creating larger passages in the formation through which water can flow more freely towards the well.

In addition to the above, well development produces two other beneficial results. Firstly, it corrects any clogging or compacting of the water-bearing formation which has occurred during drilling. Clogging is particularly evident in wells drilled by the rotary method where the drilling mud effectively seals the face of the borehole. Driving casing in the cable-tool percussion method vibrates the unconsolidated particles, thus compacting them. These are not the only drilling methods that damage the formation in one way or the other. All drilling methods do to different degrees of magnitude, and well development is needed to correct this damage.

Secondly, well development grades the material in the water-bearing formation immediately around the screen in such a way that a stable condition in which the well yields sand-free water at maximum capacity is achieved. In a zone just outside the screen, all particles smaller than the size of the screen openings are removed by development, thus leaving only the coarsest material in place. A little farther away some medium-sized grains remain mixed with the coarser ones. This grading of coarse through successively less coarse material continues as distance from the screen increases until material of the original character of the water-bearing formation is reached. This marks the end of the developed zone around the well. The succession of graded zones of material around the screen stabilizes the formation so that no further sand movement will take place. The extent of the envelope depends upon the formation characteristics, the well screen design and the skill of the well driller. Fig. 6.1 illustrates the principle of well development described above and which applies to naturally developed wells. Gravel-packed wells present a somewhat different problem which is discussed later in the chapter

Fig. 6.1 HIGHLY PERMEABLE DEVELOPED ZONE AROUND WELL SCREEN. ALL MATERIAL FINER THAN THE SCREEN OPENINGS HAS BEEN REMOVED. REMAINING MATERIAL GRADED FROM COARSER TO FINER SIZES WITH DISTANCE FROM THE SCREEN.

The development operation, to be effective, must cause reversals of flow through the screen openings and the formation immediately around the well (Fig. 6.2). This is necessary to avoid the bridging of openings by groups of

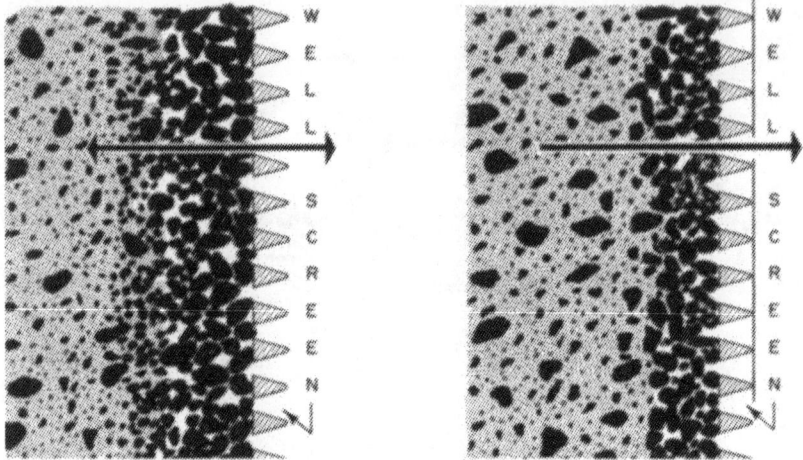

Fig. 6.2   EFFECTIVE DEVELOPMENT REQUIRES MOVING WATER IN AND OUT THROUGH SCREEN OPENINGS TO BREAK DOWN SAND BRIDGING CAUSED BY WATER MOVEMENT IN ONE DIRECTION ONLY.

particles as can occur when flow is continuously in one direction. The bridging effect of one-directional flow is illustrated in Fig. 6.3. The reversals

Fig. 6.3   ONE DIRECTIONAL FLOW CAN CAUSE SAND BRIDG-ING DURING WELL DEVEL-OPMENT.

of flow are caused by forcing water out of the well through the screen and into the water-bearing formation and then re-moving the force to allow flow to take place from the formation through the screen and back into the well. This process is known as *surging*. The outflow (with respect to the well) portion of the surge cycle breaks down any bridging of openings that may occur while the inflow portion moves the fine ma-terial toward and through the screen into the well from which it is later removed. There are several methods of producing the desirable surging action to develop a well. Some of the simpler of these are described in the paragraphs that follow.

Mechanical Surging

Mechanical surging is the name given to the method by which a plunger is operated up and down in the casing like a piston in a cylinder. The tool normally used is called a *surge plunger* or *surge block*. It is the most widely used tool for well development. Surge plungers are of two types, the solid plungers and valve-type plungers.

98

A *solid-type surge plunger* is shown in Fig. 6.4. It is of simple construction, consisting of two leather or rubber-belt discs sandwiched between wooden discs, all assembled over a pipe nipple with steel plates serving as washers under the end couplings. The leather or rubber discs should form a reasonably close fit in the well casing. This is by no means the only way of making a solid-type surge plunger. It is only one of several ways of so doing but serves to illustrate the essential features of this tool. Variations could include the use of cupped leather or rubber facing on the wooden discs instead of the flat leather or rubber-belt discs. A simple form of plunger can also be made for use in small diameter wells by securely tying enough strips of sacking around the drill pipe (preferably at a joint) to obtain a close fit in the well casing.

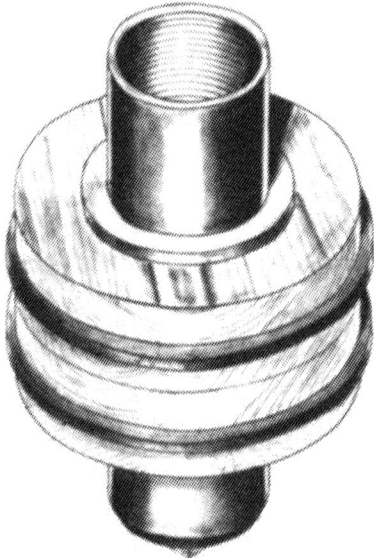

Fig. 6.4 TYPICAL SOLID-TYPE SURGE PLUNGER.

Before surging, the well should be washed with a jet of water and bailed or pumped to remove some of the mud cake on the face of the borehole and any sand that may have settled in the screen. This ensures that a sufficiently free flow of water will take place from the aquifer into the well to permit the plunger to run smoothly and freely. The surge plunger is then lowered into the well (Fig. 6.5) to a depth 10 to 15 feet under the water but above the top of the screen. A spudding motion is then applied, repeatedly raising and dropping the plunger through a distance of 2 to 3 feet. If a cable-tool drilling rig is used, it should be operated on the long-stroke spudding motion. It is important that enough weight be attached to the surge plunger to make it drop readily on the downstroke. A drill stem or heavy string of pipe is usually found adequate for this purpose.

Surging should be started slowly, gradually increasing the speed but keeping within the limit at which the plunger will rise and fall smoothly. Surge for several minutes, noting the speed, stroke and time for this initial operation. Withdraw the plunger, lower the bailer or sand pump into the well and after checking the depth of sand accumulated in the screen, bail the sand out. Repeat the surging operation, comparing the quantity of sand with that brought in initially. Bail out the sand and repeat the surging and bailing operaitons until little or no sand is pulled into the well. The time should be increased for each successive period of surging as the rate of entry of sand into the well decreases. The sand-pump type of bailer described earlier in Chapter 5 is generally favored for removing sand during development work.

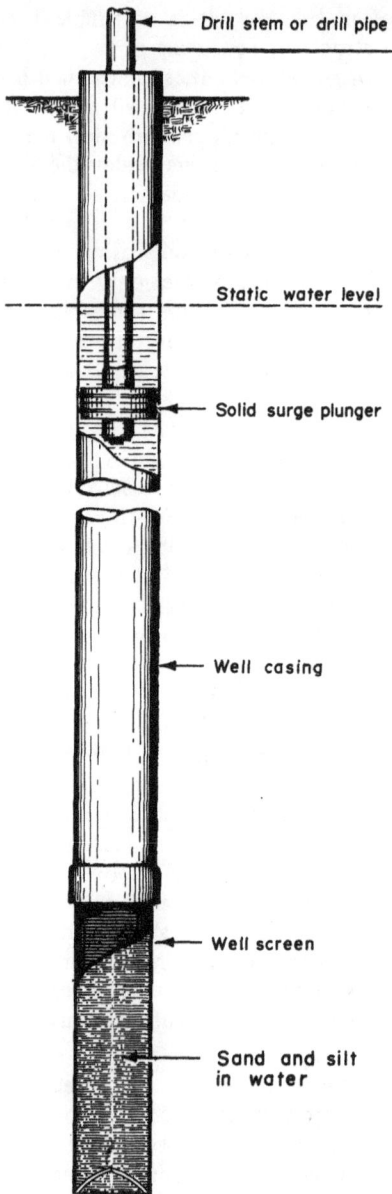

Drill stem or drill pipe

Static water level

Solid surge plunger

Well casing

Well screen

Sand and silt in water

Fig. 6.5 SOLID-TYPE SURGE PLUNG-ER READY FOR USE IN DE-VELOPING A WELL. DOWN-STROKE FORCES WATER OUTWARD INTO SAND FOR-MATION. UPSTROKE PULLS IN WATER, SILT AND FINE SAND THROUGH SCREEN.

The *valve-type surge plunger* differs from the solid-type surge plunger in that the former carries a number of small portholes through the plunger which are covered by soft valve leather. In Fig. 6.6 the valve leather is raised to indicate one of the six portholes which are spaced at equal distances around the circumference of the plunger.

Valve-type surge plungers are operated in a similar manner to solid plungers. They pull water from the aquifer into the well on the upstroke and, by allowing some of the water in the well to press upward through the valves on the downstroke, produce a smaller reverse flow in the aquifer. This creation of a greater in-rush of water to the well than out-rush during the surging operation is the principal and most important feature of this type of plunger. The valve-type surge plunger, because of this feature, is particularly suited to use in developing wells in formations with low permeabilities, since it ensures a net flow of water into the well rather than out of it. A net outward flow can result in the water moving upwards to wash around the outside of the casing since the low permeability of the aquifer will not permit flow readily into it. Washing around the outside of the casing could cause caving of the upper formations and thus create very difficult problems.

An incidental benefit gained from the use of this type of plunger is the accumulation of water above the plunger with the eventual discharge of some water, silt and sand over the top of the well. The valves in effect produce a sort of pumping action in addition to the surging of

Fig. 6.6 **TYPICAL VALVE-TYPE SURGE PLUNGER WITH VALVE LEATHER RAISED TO SHOW ONE OF SIX PORT-HOLES.**

the well and thus reduce the number of times it is necessary to remove the plunger to bail sand out of the well.

Surge plungers can also be operated within the screen. This may be desirable in developing wells with long screens. By operating the plunger within the screen, the surging action can be concentrated at chosen levels until the well is fully developed throughout the entire length of the screen. The surge plungers should, for such use, be sized to pass freely through the screen and its fittings and not form a close fit in them, as is the case when operating within the well casing. Special care must be exercised when surging within the screen to prevent the plunger from becoming sand-locked by the settling of sand above it. For this reason the use of plungers within screens should only be attempted by experienced drillers.

Care must also be exercised when using surge plungers to develop wells in aquifers containing many clay streaks or clay balls. The action of the plunger can, under such conditions, cause the clay to plaster over the screen surface with a consequent reduction rather than increase in yield. In addition, surging of the partly or wholly plugged screen can produce high differential pressures, with a resulting collapse of the screen.

Backwashing

High-velocity jetting or the backwashing of an aquifer with high velocity jets of water directed horizontally through the screen openings is generally the most effective method of well development. The principal items of equipment required are a simple jetting tool, a high pressure pump, the necessary hose, piping, swivel and water tank or other source of water supply.

A simple form of jetting tool for use in small wells is shown in Fig. 6.7. An appropriately sized coupling with a steel plate welded over one end is screwed to a 1-, 1-1/2- or 2-inch pipe. Two to four 3/16- or 1/4-inch diameter holes, equally spaced around the circumference are drilled through the full thicknesses of the coupling and the jetting pipe at a fixed distance along the coupling from the near surface of the steel plate. Better results would be obtained if properly shaped nozzles are used instead of the straight drilled holes shown but the latter are acceptably effective. Any of the above mentioned sizes would be suitable for use in a 4-inch well but the 2-inch or 1-1/2-inch tool would be preferable. The 1-1/2-inch tool can also be used in a 3-inch well while the 1-inch tool is recommended for use in a 2-inch well.

101

The procedure is to lower the tool on the jetting pipe to a point near the bottom of the screen. The upper end of the pipe is connected through a swivel and hose to the discharge end of a high pressure pump such as the mud pump used for hydraulic rotary drilling. The pump should be capable of operating at a pressure of at least 100 pounds per square inch (psi) and preferably at about 150 psi while delivering 10 to 12 gallons per minute (gpm) for each 3/16-inch nozzle or 16 to 20 gpm for each 1/4-inch nozzle on the tool. For example, a tool with two 3/16-inch diameter nozzles would require a pumping rate of about 20 to 24 gpm, while a tool with three 1/4-inch diameter nozzles would require a pumping rate of 48 to 60 gpm. While pumping water through the nozzles and screen into the formation, the jetting tool is slowly rotated, thus washing and developing the formation near the bottom of the well screen. The jetting tool is then raised at intervals of a few inches and the process repeated until the entire length of screen has been backwashed and fully developed.

Where possible, it is very desirable to pump the well at the same time as the jetting operation is in progress. This may be done in a 4-inch well if a 1-1/2-inch jetting pipe is used, thus permitting a small suction pipe to be lowered along side of it in the well. The static water level must be near enough to the surface to permit pumping by suction lift. By pumping more water out of the well than is added by jetting, flow will be induced into the well from the aquifer, thus bringing the formation material, loosened by the jetting, into the well and out of it with the discharged water. This speeds up the development process and makes it more efficient.

The high-velocity jetting method is more effective in wells constructed with continuous-slot type well screens. The greater percentage of open area of this type of screen permits a more effective use of the energy of the jet in disturbing and loosening formation material rather than in being dissipated by merely impinging upon the solid areas of slotted pipe (Fig. 6.8).

Jetting is the most effective of development methods because the energy of the jets is concentrated over small areas at any particular

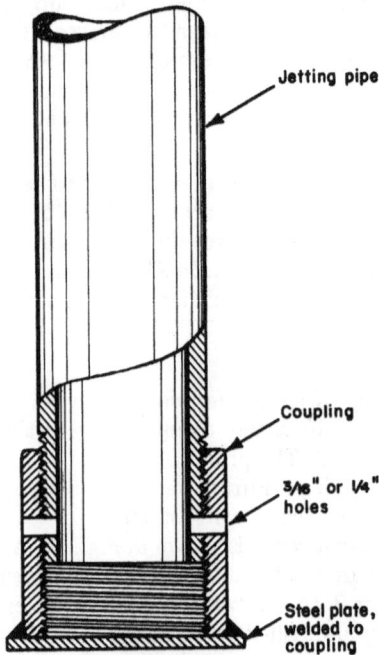

Fig. 6.7 SIMPLE TOOL FOR DEVEL-OPING WELL BY HIGH-VELOCITY JETTING METH-OD. (Adapted from Fig. 96, *Wells*, Department of the Army Technical Manual TM5-297, 1957.)

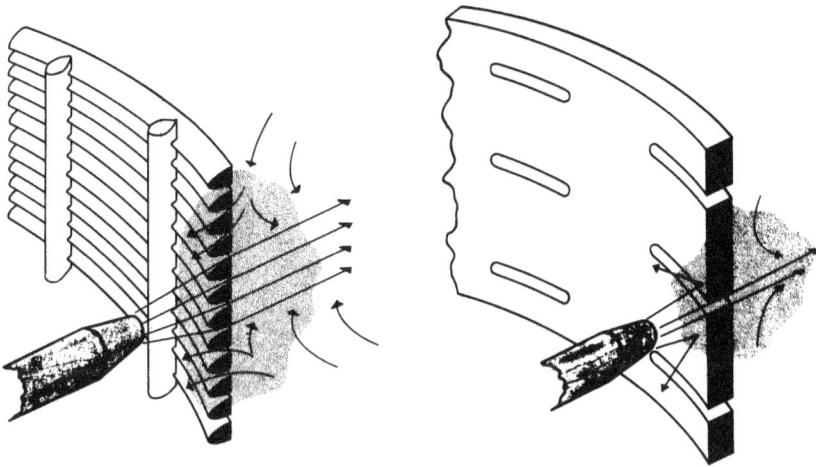

Fig. 6.8 GREATER PERCENTAGE OF OPEN AREA IN CONTINUOUS-SLOT SCREENS PERMITS BETTER DEVELOPMENT BY HIGH-VELOCITY JETTING THAN IS POSSIBLE WITH SLOTTED PIPE.

time and every part of the screen can be selectively treated. Thus uniform and complete development is achieved throughout the length of the screen. The method is also relatively simple to apply and not too likely to cause trouble as a result of over-application.

Another backwashing method of development suitable for use in small wells is one which uses a centrifugal pump with the suction hose connected directly to the top of the well casing and carrying a gate valve on the discharge end. The procedure simply involves the periodic opening and closing of the discharge valve while the pump is in operation. This creates a surging effect on the well. The process is continued until the discharge is clear and sand-free. The method is only applicable where static water levels are such as to permit pumping by suction lift. Some damage can be caused to the pump through the wearing of its parts by the sand pumped through it, particularly if in large quantities. The use of the pump to be permanently installed at the well is, therefore, not recommended for use in development of a well by this method.

Development of Gravel-Packed Wells

Development of gravel-packed wells is aimed at removing the thin skin of relatively impervious material which is plastered on the wall of the hole and sandwiched between the natural water-bearing formation and the artificially placed gravel.

The presence of the gravel envelope creates some difficulty in accomplishing the job. Success depends upon the grading of the gravel, the method of development and the avoidance of an excess thickness of gravel pack. The jetting method, because of its concentration of energy over smaller areas, is usually more effective than the other methods in developing gravel-packed

103

wells. The thinner the gravel pack, the more likely is the removal of all of the undesirable material, including any fine sand and silt. The use of dispersing agents (described immediately below) such as polyphosphates effectively assist in loosening silt and clay.

Dispersing Agents
Dispersing agents, mainly polyphosphates, are added to the drilling fluid, backwashing or jetting water, or water standing in the well to counteract the tendency of mud to stick to sand grains. These agents act by destroying the gel-like properties of the drilling mud and dispersing the clay particles, thus making their removal easier. Sodium hexametaphosphate is probably the best known of these chemical agents, though tetra sodium pyrophosphate, sodium tripolyphosphate and sodium septaphosphate are also effectively used in well development. These agents work effectively when applied at the rate of half a pound of the chemical to every 100 gallons of water in the well. The mixture should be allowed to stand for about one hour before starting development operations.

WELL DISINFECTION
Disinfection is the final step in the completion of a well. Its aim is the destruction of all disease-producing organisms introduced into the well during the various construction operations. Entry of these organisms into the well can occur through contaminated drilling water, on equipment, materials or through surface drainage into the well. All newly constructed wells with the possible exception of flowing artesian wells should, therefore, be disinfected. Wells should also be disinfected after repair and before being returned to use. The water from flowing artesian wells is generally free from contamination by disease-producing organisms after being allowed to flow to waste for a short while. If, however, analyses show persistent contamination, then the well should be disinfected as described later in this chapter.

Because of the problems of testing for specific disease-producing organisms, of which there may be several types present in water, coliform bacteria are used as indicators of the possible presence of disease-producing organisms of human or animal origin in water. Disinfection is, therefore, considered complete when sampling and testing of water show the presence of no coliform bacteria. Sampling and testing should be undertaken by experienced personnel from a health agency or recognized laboratory.

The well should be cleaned, as thoroughly as possible, of foreign substances such as soil, grease and oil before disinfection. Disinfection is most conveniently achieved by the addition of a strong solution of chlorine to the well. The contents of the well should then be thoroughly agitated and allowed to stand for several hours and preferably overnight. Care should also be taken to wash all surfaces above the water level in the well with the disinfecting solution. Following this, the well should be pumped long enough to change its contents several times and so flush the excess chlorine out of it.

Calcium hypochlorite is the most popular source of chlorine used in the disinfection of wells. It is sold in chemical supply and some hardware stores in the granular and tablet form containing 70 percent of available chlorine by weight. It is fairly stable when dry, retaining 90 percent of its original chlorine content after one year's storage. When moist, it loses its strength and becomes quite corrosive. It should, therefore, be stored under cool, dry conditions. Enough calcium hypochlorite should be added to the water standing in the well to produce a solution of strength ranging from 50 to 200 parts per million (ppm) by weight and usually about 100 ppm. A solution of approximately 100 ppm chlorine can be obtained by adding 2 ounces or 4 heaped tablespoons of calcium hypochlorite (containing 70 percent of available chlorine) to every 100 gallons of water standing in the well. Usually for convenience of application, a stock solution is made by mixing the calcium hypochlorite with a small amount of water to form a smooth paste and then adding the remainder of 2 quarts of water for every ounce of the chemical. Stir the mixture thoroughly for 10 to 15 minutes before allowing to settle. The clearer liquid is then poured off for use in the well. A gallon of this solution, when added to 100 gallons of water in the well, produces a solution of strength approximately equal to 100 ppm of chlorine. The stock solution should be prepared in a thoroughly cleaned glass, crockery or rubber lined container. Metal containers become corroded and should be avoided. Stock solutions should be prepared to meet immediate needs only since they lose strength rapidly unless properly stored in tightly covered dark glass or plastic containers. Storage of the chemical in the dry form is much more desirable.

Sodium hypochlorite may be used in the absence of calcium hypochlorite. This chemical is available only in liquid form and can be bought in strengths of up to about 20 percent available chlorine. In its most common form, household laundry bleach, it has a strength of about 5 percent of available chlorine. A stock solution of equivalent strengh to that made from calcium hypochlorite and described in the previous paragraph can be made by diluting commercial bleach with twice as much water. This stock solution should also be added to the well at the rate of one gallon to every 100 gallons of water in the well.

Flowing artesian wells are disinfected, when necessary, by lowering a perforated container, such as a short length of tubing capped at both ends, filled with an adequate quantity of dry calcium hypochlorite to the bottom of the well. The natural up flow of water in the well will distribute the dissolved chlorine throughout the full depth of the well. A stuffing box can be used at the top of the well to partially or completely restrict the flow and so reduce the chlorine losses.

105

CHAPTER 7

WELL MAINTENANCE AND REHABILITATION

Wells, like all other engineering structures, need regular, routine maintenance in the interest of a continuous high level of performance and a maximum useful life. The general tendency towards the maintenance of wells is one that can best be described as "out of sight – out of mind." Consequently, very little or no attention is paid to wells after completion until problems reach crisis levels, often resulting in the complete loss of the well. The importance of a routine maintenance program to the prevention, early detection and correction of problems that reduce well performance and useful life cannot be over-emphasized. A routine maintenance program can pay handsome dividends to a well owner and will certainly result in long-term benefits that exceed its cost of implementation.

FACTORS AFFECTING THE MAINTENANCE OF WELL PERFORMANCE

The factors affecting the maintenance of well performance or yield are numerous. Care should be taken to differentiate between those factors associated with the normal wearing of pump parts and those directly associated with changing conditions in and around the well. A perfectly functioning well, for example, can show a reduced yield because of a reduction in the capacity of the pump due to excessively worn parts. On the other hand, the excessive wearing of pump parts may be due to the pumping of sand entering the well through a corroded well screen. It is also possible for corrosion to affect only the pump, reducing its capacity, but to have little or no effect on a properly designed well.

The hydrologic conditions of some aquifers are such that the static water level drops gradually when wells are pumped continuously. While this results in reduced yields unless pumping levels are also correspondingly lowered, it is not an indication of a failure of the well itself, necessitating repairs or treatment of any form.

Most commonly, a decrease in the capacity of a well results from the clogging of the well screen openings and the water-bearing formation immediately around the well screen by incrusting deposits. These incrusting deposits (Fig. 7.1) may be of the hard cement-like form typical of the carbonates and sulfates of calcium and magnesium, the soft sludge-like forms of the iron and manganese hydroxides or the gelatinous slimes of iron bacteria. Iron may also be deposited in the form of ferric oxide with a reddish-brown, scale-like appearance. Less common is the deposition of soil materials such as silt and clay.

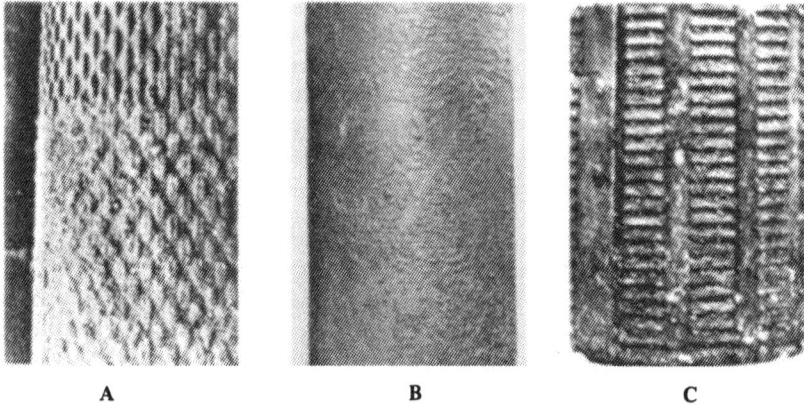

| A | B | C |

Fig. 7.1   FORMS OF INCRUSTATION.
      A. Hard cement-like deposits of calcium and magnesium carbonates.
      B. Gelatinous slime deposits typical of iron bacteria.
      C. Scale-like deposits of iron oxide completely plugging screen openings.

The deposition of carbonates and the compounds of iron and manganese can often be traced to the release of carbon dioxide from the water. The capacity of water to hold carbon dioxide varies directly with the pressure — the higher the pressure, the greater the quantity of carbon dioxide held. Pumping of a well reduces the pressure in and near the well, thus allowing the escape of carbon dioxide to the atmosphere and altering the chemical quality of the water in such a manner as to cause the precipitation of carbonate and iron deposits.

A change in velocity is another factor that can result in the precipitation of iron and manganese hydroxides. This too occurs at and near the well screen where the velocity of the slowly flowing water is suddenly increased on entry to the well.

PLANNING

The planning of well maintenance procedures should be based on a system of good record keeping. The preceding paragraphs have indicated that the problems that result in reduced well yields occur at and around the well screen and very much out of sight. The analysis of good records must, therefore, be relied upon as the source of problem detection in wells. There can be no substitute for the keeping of good records.

Among the records kept should be pumping rates, drawdown, total hours of operation, power consumption and water quality analyses. Pumping rates and drawdown are particularly useful in determining the specific capacity (discharge per foot of drawdown) which is the best indicator of existing problems in a well. The specific capacities of wells should be checked periodically and compared with previous values including those immediately after completion of the wells to determine whether significant reductions have taken place. A significant reduction in the specific capacity of a well could often be traced to blockage of the well screen and the formation around it, most likely by incrusting deposits. As stated earlier, a reduction in the pump discharge would not by itself be evidence of a reduced capacity of the well. If, however, the drawdown in the well does not show an equal reduction, then the specific capacity will be reduced, thus indicating the probability of an incrustation problem.

Power consumption records also provide valuable evidence of the existence of problems in wells. Should there be an increase in power consumption, not accompanied by a corresponding increase in the quantity of water pumped, then a problem is possible in either the pump or the well. Should an investigation show no problems in the pump nor appreciable increase in the dynamic head against which the pump has been operating, then it is most likely that a problem exists in the well and that the problem is causing an increased drawdown. A check on the drawdown should then be undertaken to verify the deduction and the well checked for incrustation.

Since there would be no incrustation in the absence of incrusting chemicals in the water, the value of chemical analyses of well water is self-evident. Such analyses are more useful as problem indicators if undertaken regularly. They indicate the type of incrustation that might occur and the expected rate of deposition in the well and its vicinity. The quality of some well waters changes slowly with time and only regular routine analyses would indicate such changes.

In wells, the waters of which are known to be incrusting, the frequency of observations of all types should be as high as possible and consistent with the use to which the water is being put. Observations should be made much more frequently at wells serving a community than at a private home-owner's well, since more people are dependent on the community wells. Power consumption, well discharge, drawdown, operating hours and other such observations are often made daily on community wells and may even be done on a continuous basis. Chemical analyses on such wells may be done on an annual, semi-annual or quarterly basis, as conditions warrant them. Observations on home-owner's wells are usually much less frequent but should, nevertheless, be undertaken regularly.

MAINTENANCE OPERATIONS

Maintenance operations should not be deferred until problems assume major proportions as rehabilitation then becomes more difficult and sometimes impossible or impracticable. Incrustation not treated early enough can so clog the well screen and the formation around it that it becomes extremely difficult and even impossible to diffuse a chemical solution to all affected points in the formation. Any attempts at rehabilitation would then prove unsuccessful.

No methods have yet been developed for the complete prevention of incrustation in wells. Various steps can be taken to delay the process and reduce the magnitude of its effects. Among these are the proper design of well screens and the reduction of pumping rates, both aimed at reducing entrance velocities into screens and drawdown in wells. For example, it may be worthwhile to share the pumping load among a larger number of wells in order to reduce the rate of incrustation. However, the ultimate or final solution will be in a regular cleaning program. Incrusting wells are usually treated with chemicals which either dissolve the incrusting deposits or loosen them from the surfaces of the well screen and formation materials so that the deposits may be easily removed by bailing.

Acid Treatment

Acid treatment refers to the treatment of a well with an acid, usually hydrochloric (muriatic) acid or sulfamic acid for the removal of incrusting deposits. Both of these acids readily dissolve calcium and magnesium carbonate, though hydrochloric acid does so at a faster rate. Strong hydrochloric acid solutions also dissolve iron and manganese hydroxides. The simultaneous use of an inhibitor serves to slow up the tendency of the acid to attack steel casing.

Wells are sometimes treated with acid in preparation for the withdrawal of a screen either for re-use elsewhere or in the same well. For example, it may be desirable to recover a screen that is in good condition from a well whose casing has been corroded beyond usefulness. Or, a screen may be recovered for more effective treatment against incrustation than can be achieved in the well. In either case, a preliminary acid treatment to dissolve some of the incrusting deposits will make it much easier to pull the screen.

Hydrochloric acid is usually available in three grades from chemical supply shops. The strongest grade, designated as the 27.92 percent grade, should be used. It is sold in either glass or plastic carboys containing about 12 gallons each. If inhibited acid cannot be obtained, unflavored gelatin added at the rate of 5 to 6 pounds to every 100 gallons of acid will prevent serious damage to steel casing.

Hydrochloric acid should be used at full strength. Each treatment usually requires 1-1/2 to 2 times the volume of water in the screen. This provides enough acid to fill the screen and additional acid to maintain adequate strength as the chemical reacts with the incrusting materials. Fig. 7.2 illustrates a method of placing acid in a well. Acid is introduced within the screen by means of a wide-mouthed funnel and 3/4- or 1-inch black iron or plastic pipe. Acid is heavier than water which it tends to displace but with which it also mixes readily to become diluted. When used in long screens, acid should be added in quantities sufficient to fill 5 feet of the screen and the conductor pipe raised 5 feet after pouring each quantity.

The acid solution in the well should be agitated by means of a surge plunger or other suitable means for 1 to 2 hours following which the well should be bailed until the water is relatively clear. The driller usually can detect an improvement in the yield of the well while running the bailer. The well may, however, be pumped to determine the extent of improvement. If this is not as expected, then the treatment may be repeated using a longer period of agitation before bailing. A third treatment may even be undertaken.

The procedure is sometimes varied to alternate acid treatment and chlorine

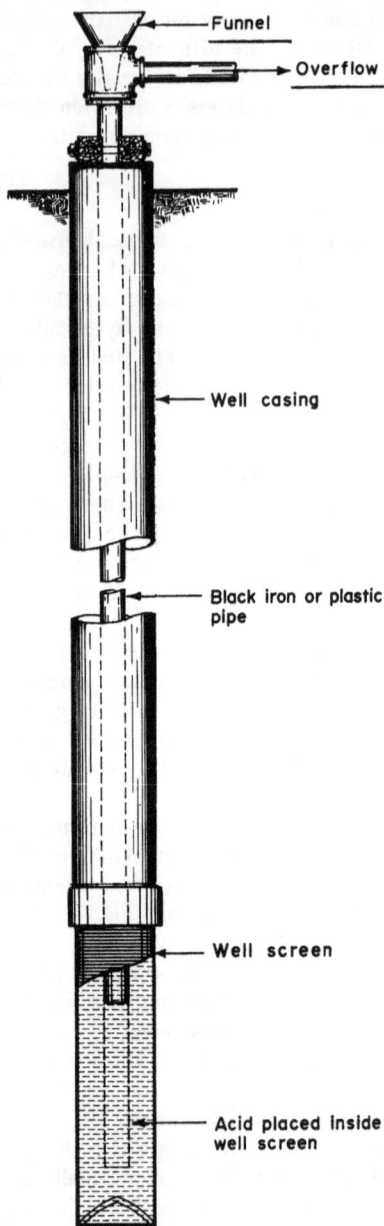

Fig. 7.2 **ARRANGEMENT FOR INTRODUCING ACID INSIDE WELL SCREEN FROM BOTTOM UPWARDS.**

Funnel
Overflow
Well casing
Black iron or plastic pipe
Well screen
Acid placed inside well screen

treatment (described later in this chapter), repeating the alternate treatments as many times as it appears that beneficial results are being obtained. The chlorine helps to remove the slime deposited by iron bacteria.

Sulfamic acid can be obtained as a dry granular material which produces a strong acid solution when dissolved in water. It offers a number of advantages over hydrochloric acid as a means of treating incrustation in wells. It can be added to a well in either its original granular form or as an acid solution mixed on site. Granular sulfamic acid is nonirritating to dry skin and its solution gives off no fumes except when reacting with incrusting materials. Spillage, therefore, presents no hazards and handling is easier, cheaper, and safer. It also has a markedly less corrosive effect on well casing and pumping equipment and is safe for use on Everdur and type 304 stainless steel well screens. These advantages tend to offset its higher cost than inhibited hydrochloric acid. Sulfamic acid dissolves calcium and magnesium carbonates to produce very soluble products. The reaction is, however, slower than that using hydrochloric acid and a somewhat longer contact period in the well is required.

Sulfamic acid is usually added to wells in solution form using a black iron or plastic pipe as described for the application of hydrochloric acid. Ten gallons of water dissolve 14 to 20 pounds of the granules depending upon the temperature of the water.

The granular material itself can, however, be poured into and mixed with the water standing in the well. The water must be agitated to en-

110

sure complete solution of the acid. The quantity of acid added in this case should be based on the total volume of water standing in the well and not on that in the screen only, as is the case if the acid is applied in solution form. An excess of the granular material may be added to keep the solution up to maximum strength while it is being used up through reaction with the incrusting material. The addition of a low-foaming, non-ionic wetting agent improves the cleansing action to some extent.

A number of *precautions* must be exercised in using any strong acid solution. Goggles and water-proof gloves should be worn by all persons handling the acid. When preparing an acid solution, always pour the acid slowly into the water. In view of the variety of gases, some of them very toxic, produced by the reaction of acid with incrusting materials, adequate ventilation should be provided in pump houses or other confined spaces around treated wells. Personnel should not be allowed to stand in a pit or depression around the well during treatment because some of the toxic gases such as hydrogen sulfide are heavier than air and will tend to settle in the lowest areas. After a well has been treated, it should be pumped to waste to ensure the complete removal of all acid before it is returned to normal service.

Chlorine Treatment
Chlorine treatment of wells has been found more effective than acid treatment in loosening bacterial growths and slime deposits which often accompany the deposition of iron oxide. Because of the very high concentrations required, 100 to 200 ppm of available chlorine, the process is often referred to as shock treatment with chlorine. Calcium or sodium hypochlorite may be used as the source of chlorine as described for the disinfection of wells in Chapter 6. The chlorine soltuion in the well must be agitated. This may be done by using the high-velocity jetting technique (see "Well Development," Chapter 6) or by surging with a surge plunger or other suitable techniques. The recirculation provided with the use of the jetting technique greatly improves the effectiveness of the treatment.

The treatment should be repeated 3 or 4 times in order to reach every part of the formation that may be affected, and it may also be alternated with acid treatment, the latter being performed first.

Dispersing Agents
Polyphosphates, or glassy phosphates as they are commonly called, effectively disperse silts, clays and the oxides and hydroxides of iron and manganese. The dispersed materials can be easily removed by pumping. In addition, the polyphosphates are safe to handle. They find considerable application, therefore, in the chemical treatment of wells.

For effective treatment, 15 to 30 pounds of polyphosphate are added to every 100 gallons of water in the well. A solution is usually made by suspending a wire basket or burlap bag containing the polyphosphate in a tank of water. About a pound of calcium hypochlorite should be added for every 100 gallons of water in the well in order to facilitate the removal of iron bacteria and their slimes and also for disinfection purposes. After pouring this polyphosphate and hypochlorite solution into the well, a surge plunger or the jetting technique is used to agitate the water in the well. The recirculation of the solution with the use of the high-velocity jetting technique greatly improves the effectiveness of the treatment. Two or more successive treatments may be used for better results.

WELL POINT INSTALLATION IN DUG WELLS

Dug wells are holes or pits dug by hand or machine tools into the ground to tap the water table. They are usually 3 to 20 feet in diameter, 10 to 40

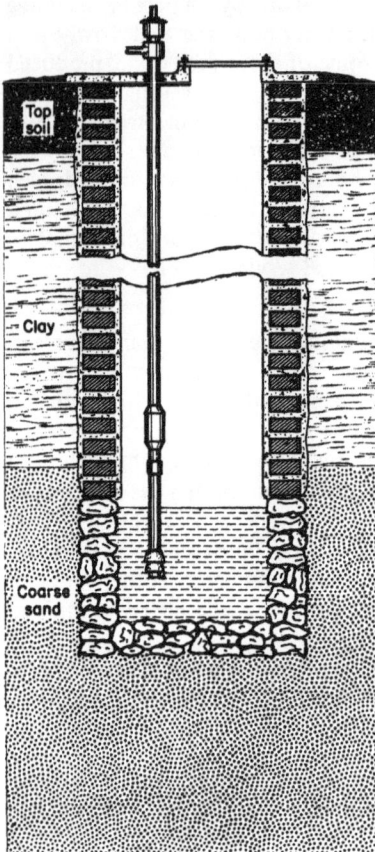

Fig. 7.3   DUG WELL.

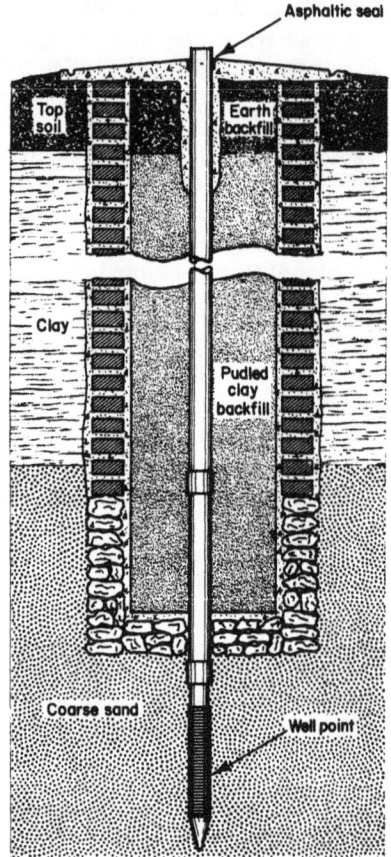

Fig. 7.4   DUG WELL OF FIG. 7.3 CONVERTED TO SAFER AND MORE PRODUCTIVE TUBULAR WELL WITH DRIVEN WELL POINT AS SCREEN.

feet deep and lined with brick, stone, tile, wood cribbing or steel rings to prevent the walls from caving (Fig. 7.3). They depend entirely on natural seepage from the penetrated portion of water-bearing formations for their yield of water.

This type of well is at a disadvantage on two scores when compared with tubular wells of the type so far described. Firstly, dug wells are much more difficult to maintain in a sanitary condition. Secondly, their yields are very low, because they do not penetrate very far into the water-bearing formation and cannot be developed in a similar manner to screened wells.

Dug wells usually can be made much safer and more productive by driving well points into the water-bearing formation and thus converting them into tubular wells. A properly developed well with a short length of 2" drive-point screen will usually produce water at a much higher rate than can be had from a dug well several feet in diameter. The annular space between the casing of the driven well and the wall of the existing well should be back-filled with a puddled clay or other suitable material. The sanitary precautions with respect to the completion of the upper terminal of a well (described in Chapter 4) should be observed. The wall of the existing dug well may be cemented prior to back-filling. A converted dug well is illustrated in Fig. 7.4.

CHAPTER 8

PUMPING EQUIPMENT

Drilling and completing a well form only part of a solution to the problem of getting water in sufficient quantity where it is desired for use. Small wells are generally used for supplying water to a home, a group of homes or other such limited consumers of water as a small factory. The water is usually required for use at elevations somewhat higher than those at which the water is found in the well and, often, some appreciable distance from the well. Therefore, some means must be found of lifting the water from a well and forcing it through pipes at suitable velocities to the points and elevations of use. The exception to this general statement is the case of the flowing artesian well, which has a sufficiently high discharge at an adequate pressure to meet the limited demands of one or a few small homes without any external help. Generally, however, help is needed, and this is provided in the form of a suitable pump. It is important that the pump be a suitable one, selected on the basis of the demand to be fulfilled and the capacity of the well to yield water. It cannot and must not be just any pump, as it is then unlikely that the needs will be met. Pump selection is discussed later in this chapter.

Pumps do not develop power of their own. Some external source of power must be provided to drive a pump and so cause it to lift and force water from one point to another. The source of power may be the man who uses his hand to operate a lever upward and downward or forward and backward or who turns a wheel connected to the pump. In this case, the pump is said to be manually operated or hand driven. The power source may also be a windmill, an electric motor or an engine which burns a fuel such as gasoline or diesel oil. A very common error is not being able to distinguish between a pump and its motive or driving force, particularly when that force is an engine or motor directly coupled to the pump. Care should be taken to avoid this, as the problems of pumps, engines and motors are very different and need different approaches to solve them.

The action of most pumps can be divided into two parts. The first is the lifting of the water from some lower level to the pump intake or suction side of the pump. The second is concerned with applying pressure to the water in the pump to force the water to its destination.

Suction lift: Consider an open-ended tube which is suspended vertically in a large container of water (Fig. 8.1A). Since the water both within and without the tube is exposed to the atmosphere, the only external force acting

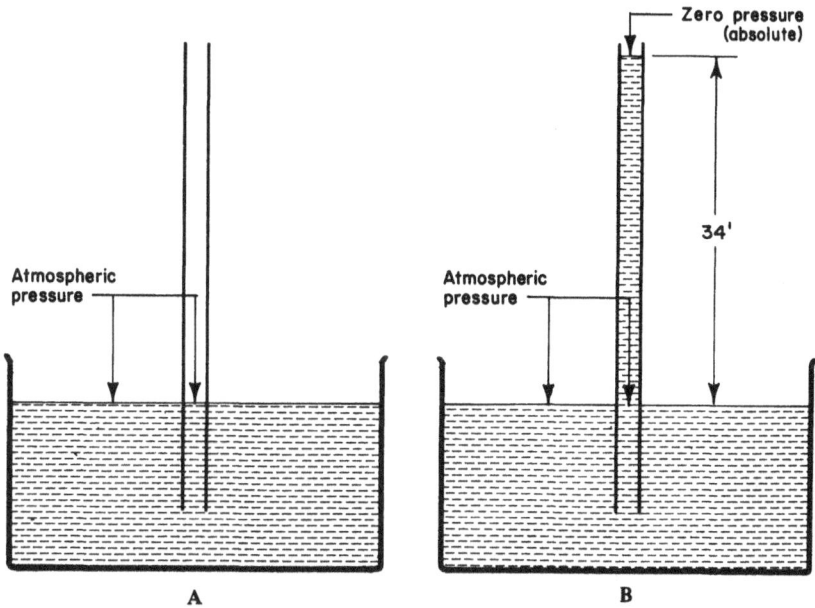

Fig. 8.1 A. ATMOSPHERIC PRESSURE THROUGHOUT. NO DIFFERENCE IN WATER LEVELS.
B. PRESSURE IN TUBE REDUCED TO ZERO ATMOSPHERES (TOTAL VACUUM). WATER LEVEL IN TUBE RISES TO APPROXIMATELY 34 FT.

on both surfaces is that due to atmospheric pressure. The pressure on the water surface being the same within and without the tube, there will be no difference in the water levels (assuming a wide enough tube that capillary forces may be neglected). If, however, the pressure on the water surface within the tube is reduced below atmospheric pressure while that outside of the tube remains at atmospheric pressure, then water will rise in the tube until the weight of the column of water inside the tube exerts a pressure equal to the original pressure difference on the water surfaces within and without the tube. The maximum height to which this column will rise occurs when the pressure on the water surface within the tube is reduced to zero atmospheres (absolute). The water column will then be exerting a downward pressure equal to the atmospheric pressure (Fig. 8.1B). Atmospheric pressure at sea level is approximately equivalent to a column of water 34 feet high, and this is the height to which the water will rise in the tube. Atmospheric pressure decreases as the altitude or height above sea level increases. Accordingly, the maximum height to which the water column can be made to rise also decreases with increase of altitude.

The term *suction* is used to describe the amount by which the pressure in the tube is reduced below atmospheric pressure. Suction can be applied to the tube by operating a pump attached to the top end of the tube. The level to which the water rises within the tube above the water surface in the large

115

container is termed the *suction lift.* A pump, in order to pump water, must be able to create enough suction to lift the water in the tube to the level of the suction end of the pump. In Fig. 8.2, the well casing represents the larger container while the suction pipe of the pump takes the place of the tube.

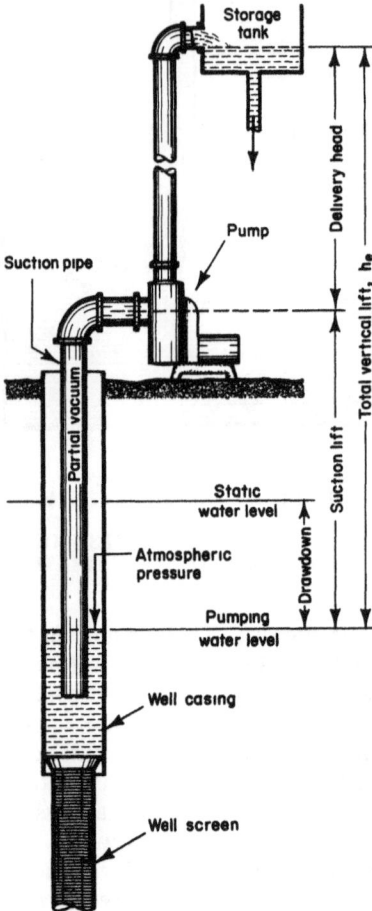

Note that the lifting of the water in the suction pipe must be accompanied by a lowering of the water level in the well casing. The water level within the casing and the suction pipe before the pump created the suction lift is called the *static water level.* The level in the well casing during pumping is the *pumping water level.*

In theory then, a pump, by creating zero pressure (absolute) or a total vacuum within its suction pipe, should be capable of a suction lift of approximately 34 feet of water at sea level and somewhat less at higher altitudes. In practice, however, this is not achieved, as pumps are not 100 percent efficient, and other factors such as water temperature and friction or resistance to flow in the suction pipe reduce the suction lift. At sea level, the best designed pumps usually achieve a suction lift of about 25 feet, while the suction lift of an average pump varies from 15 to about 18 feet. Should it be necessary to lift water in a well from a level 25 feet or more below ground surface, some means must be found of lowering the pump into the well and either completely submerging the pump in the water or taking it near enough to the water surface to permit suction lifting of the water.

Fig. 8.2 PRINCIPLES OF PUMPING A WATER WELL.

This limiting suction lift is used to classify pumps into surface-type or shallow well pumps and deep well pumps. *Surface-type pumps* are those pumps which are placed at or above ground surface and are limited to lifting water by suction from a depth usually no greater than about 25 feet below the ground surface. *Deep well pumps* are those pumps which are placed within the well and are used for extracting water from depths generally in excess of 25 feet below the ground surface.

116

Another very common classification of pumps divides them into two main types based on the mechanical principles involved. These two types are *constant displacement* and *variable displacement* pumps.

CONSTANT DISPLACEMENT PUMPS

Constant displacement pumps are so designed that they deliver substantially the same quantity of water regardless of the pressure head against which they are operating. That is to say, the rate of discharge is essentially the same at low or high pressures. However, the input power or driving force varies directly in proportion to the pressure in the system and must be doubled if the pressure is doubled. There are three main designs of this type of pump which are commonly used in water wells. These are *reciprocating piston* pumps, *rotary* pumps and *helical rotor* pumps.

Reciprocating Piston Pumps

Reciprocating piston pumps, the most common type of constant displacement pump, use the up and down or forward and backward (reciprocating) movement of a piston or plunger to displace water in a cylinder. The flow in and out of the cylinder is controlled by valves. The basic principles and steps in the operation of a single-acting piston pump are illustrated in Fig. 8.3. The

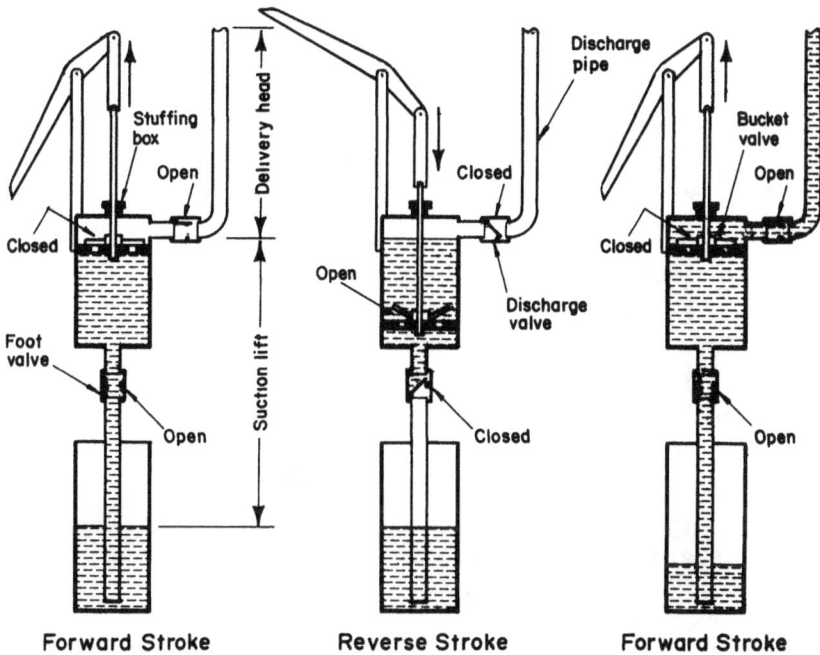

Fig. 8.3 **PRINCIPLES OF A SINGLE-ACTING RECIPROCATING PISTON PUMP.**
(Adapted from Fig. 38, *Water Supply For Rural Areas And Small Communities,* WHO Monograph Series No 42, 1959.)

117

plunger on the forward stroke pushes water from the cylinder through the open discharge valve into the discharge pipe while at the same time creating a suction behind it that opens the foot valve and permits water to flow through the suction pipe into the cylinder. The reverse stroke creates a pressure behind the piston in the cylinder, thus closing the foot valve and opening the bucket valves in the piston to let water through to the discharge side of the piston. Continuous repetition of the forward and reverse strokes of the piston results in a steady flow of water out of the discharge pipe. The amount of pressure developed by such a pump depends upon the power applied in operating the piston. These pumps are manufactured in both the surface (Fig. 8.4) and deep well (Fig. 8.5) types and may be manually or engine operated. A manually operated, surface-type piston pump is commonly known as a pitcher pump.

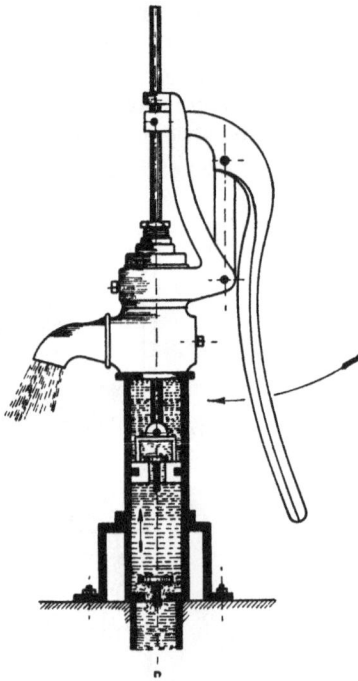

Fig. 8.4   MANUALLY OPERATED SURFACE-TYPE PISTON PUMP. (From Fig. 39, *Water Supply For Rural Areas And Small Communities*, WHO Monograph Series No. 42, 1959.)

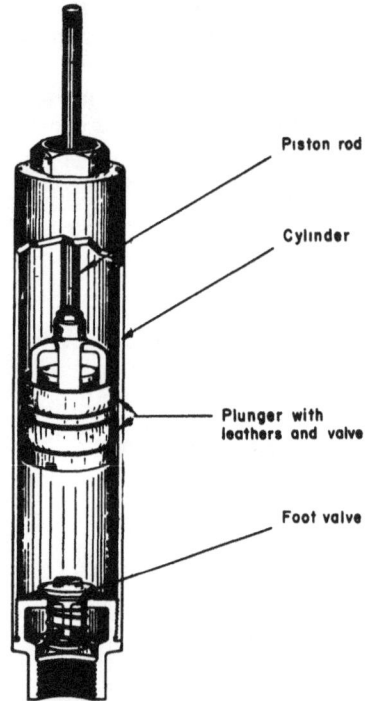

Fig. 8.5   DEEP WELL SINGLE-ACTING PISTON PUMP.

The basic principle of the single-acting piston pump can be modified to cause water to be pumped on both the forward and the reverse strokes. Pumps thus modified are known as double-acting piston pumps. Other modifications include the use of two or more pistons so that a continuous stream of water is pumped with minimum pulsation against high pressures.

The delivery or discharge rate from reciprocating piston pumps is determined by multiplying the volume of water displaced in the cylinder on each stroke by the number of strokes of the piston in a given time. Thus the discharge rate can be varied within wide limits by varying the speed of the piston. Only when slipping (movement of the water backwards between the piston and the cylinder walls) results from too rapid motion of the piston is the limit of capacity reached. This, of course, assumes an adequate power supply. There can, therefore, be great flexibility in the use of this type of pump to meet varying water demands. Other advantages are their low initial cost, sturdy construction and ease of maintenance which is normally confined to the replacement of piston washers.

Rotary Pumps

Rotary pumps generally use a system of rotating gears (Fig. 8.6) or vanes to create a suction at their inlet sides and force a continuous stream of water out of their discharge sides. They are usually surface-type pumps with capacities governed by the speed and width of the gear teeth or vanes and, when used on wells, limited by their suction lift. Gear pumps are intended for low speed operation and for the pumping of clean water, free from sand or grit since these materials can cause considerable wear on the closely fitting gear teeth.

Semi-rotary hand operated pumps of the double-acting or quadruple-acting type are commonly used in individual water supply systems in rural areas for low lifts of water from wells to overhead tanks. Fig. 8.7 illustrates the operation of a typical double-acting, semi-rotary pump. These pumps are only capable of very small suction lifts when not fitted with foot valves. However, when provided with such valves, they can operate with suction lifts of up to 20 feet.

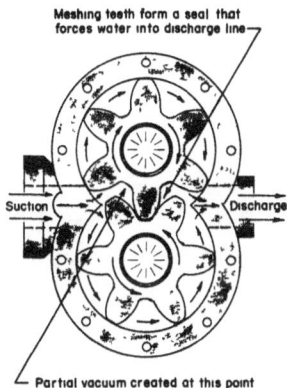

Fig. 8.6 ROTARY GEAR PUMP. (Adapted from Fig. 111, *Wells*, Department of the Army Technical Manual TM5-297, 1957.)

Helical Rotor Pumps

The helical rotor or screw-type pump is a modification of the rotary type of constant displacement pump. The main elements of the pump are the highly polished metal rotor or screw in the form of a helical, single thread worm and the outer stator made of rubber. Flexible mountings allow the rotor to rotate eccentrically within the stator, pressing a continuous stream of water forward along the cavities in the stator. The water also acts as a lubricant between the two elements of the pump. Helical rotor pumps can be either of the surface

119

or deep well type and are usually driven by engines or electric motors. Fig. 8.8 illustrates a deep well, helical rotor pump.

VARIABLE DISPLACEMENT PUMPS

The distinguishing characteristic of variable displacement pumps is the inverse relationship between their discharge rates and the pressure heads against which they pump. That is to say, the pumping rate decreases as the pressure head increases. The opposite is also true, the pumping rate increases as the pressure head reduces. The two main types of variable displacement pumps used in small wells are *centrifugal* and *jet* pumps.

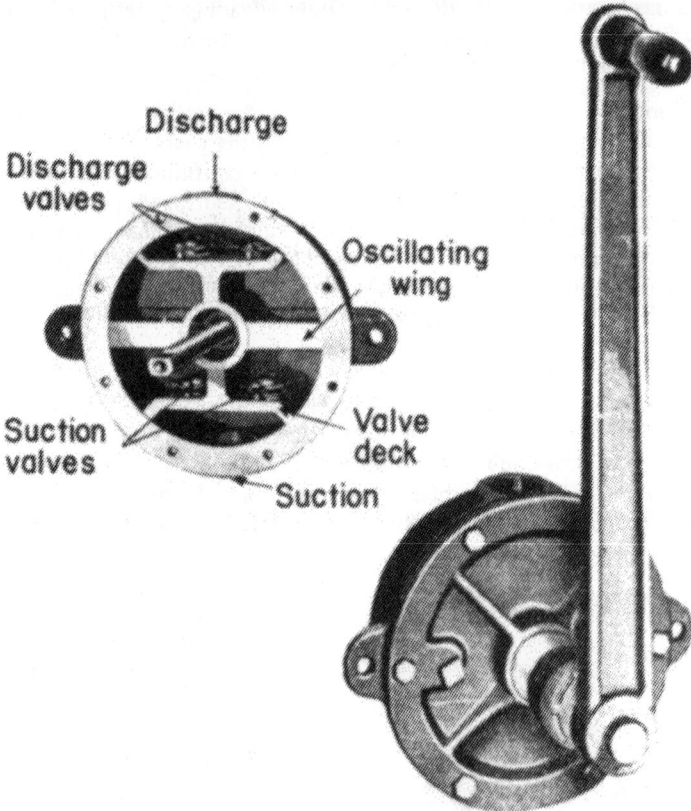

Fig. 8.7 **DOUBLE-ACTING, SEMI-ROTARY HAND PUMP.** (From Deming Division of Crane Company, Salem, Ohio.)

Centrifugal Pumps

Centrifugal pumps are the most common types of pumps in general use. The basic principles of their operation can be illustrated by considering the effect of swinging a pail of water around in a circle at the end of a rope. The force that causes the water to press outward against the bottom of the pail rather than run out at the open end is known as the centrifugal force. If a hole were cut in the bottom of the pail, water would be discharged through the opening at a velocity which is related to the centrifugal force. Further, should an intake pipe be connected to an air-tight cover on the pail, a partial vacuum would be created inside the pail as water is discharged. This vacuum could bring additional water into the pail from a supply placed at the other end of the intake pipe within the limit of the suction lift created by the vacuum. Thus a continuous flow of water could be maintained in a manner similar to that operating in a centrifugal pump. The pail and cover correspond to the casing of the pump, the discharge hole and intake pipe correspond to the pump outlet and intake respectively, while the rope and arm perform the functions of the pump impeller.

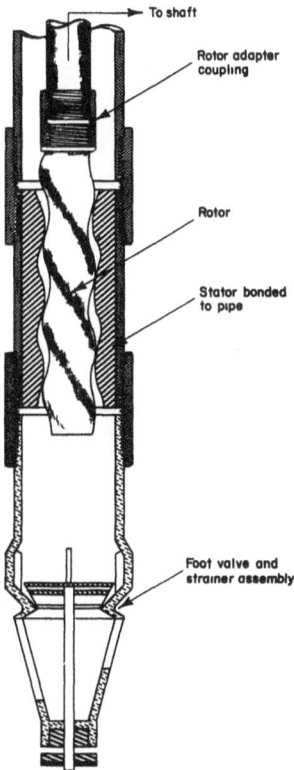

Fig. 8.8 DEEP WELL HELICAL ROTOR PUMP.

Centrifugal pumps used on small wells can be sub-divided into two main types based on their design features. These are *volute* pumps and *turbine* or diffuser pumps. The impellers of volute pumps are housed in spirally shaped casings (Fig. 8.9) in which the velocity of the water is reduced upon leaving the impeller with a resulting increase in pressure. In turbine pumps the impellers are surrounded by diffuser vanes (Fig. 8.10). These vanes provide gradually enlarging passages through which the velocity of the water leaving the impeller is gradually reduced, thus transforming the velocity head into pressure head.

The conditions of use determine the choice between volute and turbine pumps. The volute design is very commonly used in surface-type pumps where pump size is not a limiting factor and design heads are low to medium. Deep well centrifugal pumps are, however, of the turbine type of design which is better suited to use where the diameter of the pump must be limited, in this case by the diameter of the well casing.

The performance of a centrifugal pump depends greatly upon the design of its impeller. For example, the pump discharge against a given head can be increased by enlarging the diameter of the inlet eye and the width of the impeller. It is also customary to use a larger number of guide vanes (up to 12) in turbine pumps when a higher pressure head is desired. The extent to which the pressure head can be increased by an increase in the number of guide vanes is, however, limited. Greater increases are achieved by the use of multiple stages, each of which contains an impeller. Multi-stage design is used in both surface-type and deep well pumps but is particularly common in deep well pumps designed for use under high lift conditions. Generally, the discharge of a multi-stage pump is almost the same as for a single stage of the same pump. The pressure head developed and horsepower required for

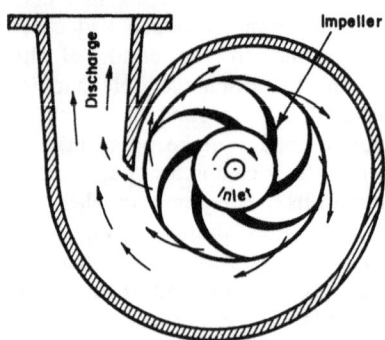

Fig. 8.9   VOLUTE-TYPE CENTRIFU-
GAL PUMP.

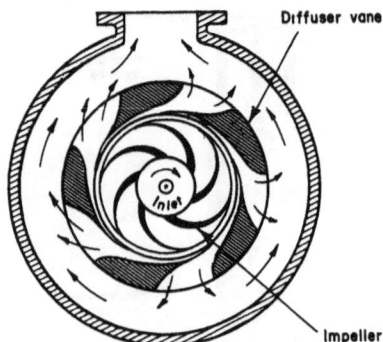

Fig. 8.10 TURBINE-TYPE CENTRIFU-
GAL PUMP SHOWING CHAR-
ACTERISTIC DIFFUSER
VANES.

operation, however, increase in direct proportion to the number of stages or impellers. For example, the pressure head of a 4-stage pump, one stage of which develops a pressure of 40 feet head of water, would be 4 times 40 or 160 feet of water. Fig. 8.11 shows a section through a 3-stage deep well turbine pump which is, in effect, three pumps assembled in series with flow passing from one to the next and the head being increased with passage through each stage.

Jet Pumps

Jet pumps, in reality, combine centrifugal pumps and ejectors to lift water from greater depths in wells than is possible through the use of surface-type centrifugal pumps when acting alone. The basic components of ejectors are a nozzle and a venturi tube shown in Fig. 8.12. The operating principles are as follows. Water under pressure is delivered by the centrifugal pump (mounted at ground level) through the nozzle of the ejector. The sudden increase in the velocity of the water as it passes through the tapered nozzle causes a reduction in pressure as the water leaves the nozzle and enters the venturi tube. The higher the water velocity through the nozzle, the greater is the reduction in pressure at the entrance to the venturi tube. This reduction in pressure can, therefore, be made great enough to create a partial vacuum and so suck water from the well through the intake pipe of the ejector and into the venturi tube. The gradual enlargement of the venturi tube reduces the velocity of flow with a minimum of turbulence and so causes a recovery of almost all of the water

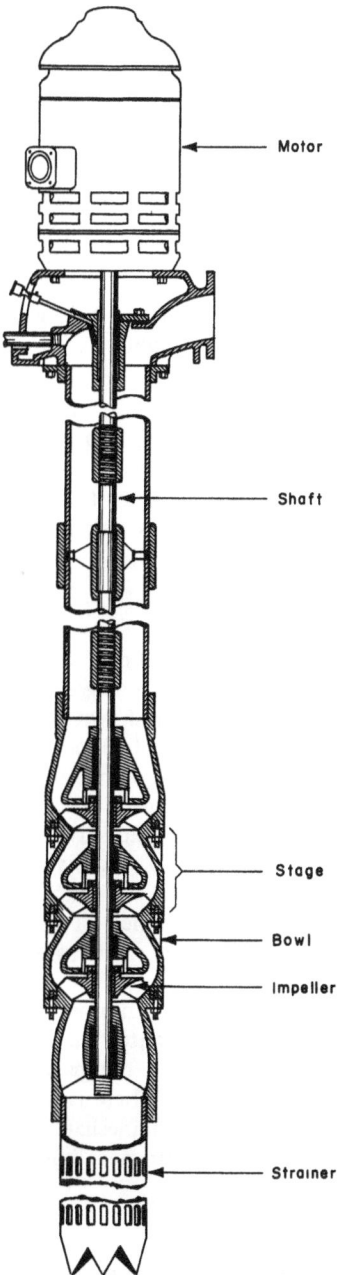

Fig. 8.11 THREE-STAGE LINESHAFT DEEP WELL TURBINE PUMP.

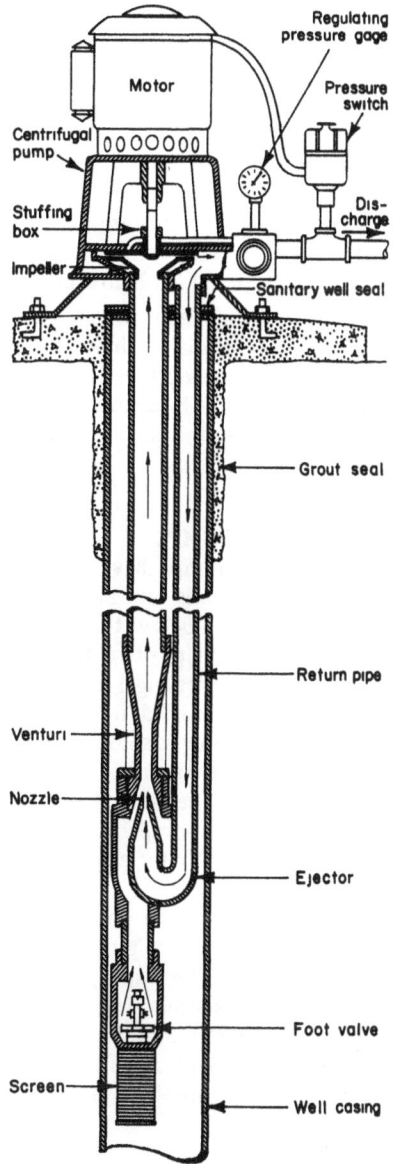

Fig. 8.12 JET PUMP. (Adapted from Fig. 13, *Manual of Individual Water Supply Systems*, Public Health Service Publication No. 24, 1962.)

pressure in going through the nozzle. The centrifugal pump then picks up the flow, sending part of it through the discharge pipe and returning the remainder to the ejector to induce more flow from the well and so repeat the cycle. The pressure regulating gage is set to maintain the necessary pressure to produce flow at the desired pumping head.

The centrifugal pump is the prime mover without which the ejector cannot pump water. Considerable increases in the discharge pressure head cannot be achieved by adjusting the regulating gage. Such pressure increases are provided by increasing the number of pump stages. Operating conditions should always be such that the ejector nozzle is covered by at least 5 feet of water. Small jet pumps are usually limited to discharges of about 20 gallons per minute against total pressure heads not exceeding about 150 feet of which the required lift below ground is about 100 feet or less.

Jet pumps are, generally, inefficient pumps but have a number of desirable features which make their use popular in small, domestic water supply installations. Among these features are their adaptability to use in small wells down to 2 inches in diameter, the accessibility of the moving parts which are all above the ground surface, their simplicity and relatively low purchase price and maintenance cost.

DEEP WELL PUMPS

Deep well pumps have been earlier defined as those pumps which are placed within wells and are used for lifting water from depths generally in excess of 25 feet below the ground surface. It has also been shown that they can be of both the positive displacement (piston and helical rotor) and variable displacement (centrifugal and jet) types of design. Deep well pumps are, however, further classified in accordance with the positioning of their power source. If the power source is situated at or above the ground surface, thus necessitating the transmission of the driving force through a long shaft down to the pump in the well, then the pump is referred to as a vertical *lineshaft* pump. Lineshaft pumps may be driven either by direct-coupled electric motors (Fig. 8.11) or by engines or electric motors through right-angle drive heads (Fig. 8.13).

When, however, the power source (in this case an electric motor) is fitted immediately below the pump and submerged with it in the water, the pump is called a *submersible* pump (Fig. 8.14). Shafts in submersible pumps only extend from the submerged motor to the top-most impeller. There is no shaft between the pump and the ground surface as is necessary in lineshaft pumps. This feature provides submersible pumps with one of their more important advantages over lineshaft pumps.

Lineshaft Pumps

Lineshaft pumps have been in use for several years, preceding their more recent competitors, submersible pumps. Most failures in pumping installations usually occur as a result of problems arising in the power source. Lineshaft pumps, by having their power sources installed above ground level and separated from the pump, make access to these power sources easier and repairs possible without removing the whole pump assembly from the well. Greater flexibility can also be achieved by the use of a dual right-angle drive head to which two engines, two electric motors or one engine and an electric motor can be coupled. This arrangement permits the use of a stand-by power

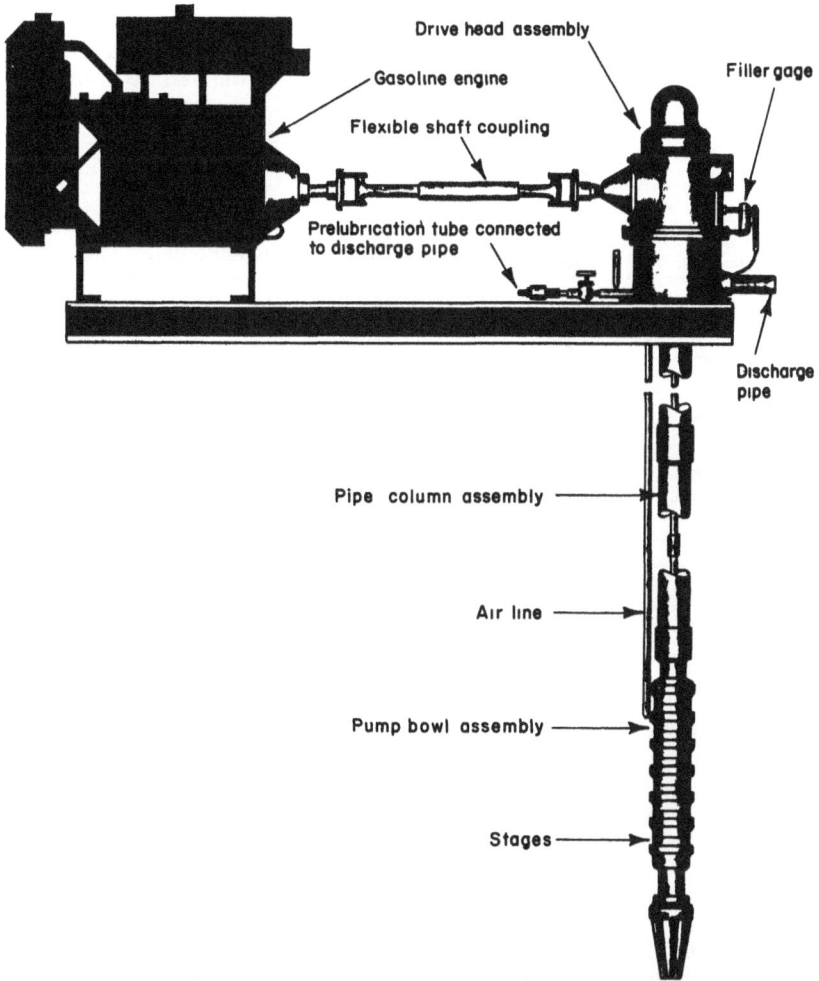

Fig. 8.13 ENGINE DRIVEN, LINESHAFT DEEP WELL TURBINE PUMP. (Adapted from Fig. 116, *Wells*, Department of the Army Technical Manual TM5-297, 1957.)

source and continuous operation of the pump by one source while the other is being serviced or repaired.

Lineshaft pump installations must, however, be enclosed in pump houses and, partly as a result of this, are usually more costly than submersible pump installations. The shafts and bearings of lineshaft pumps also provide many more

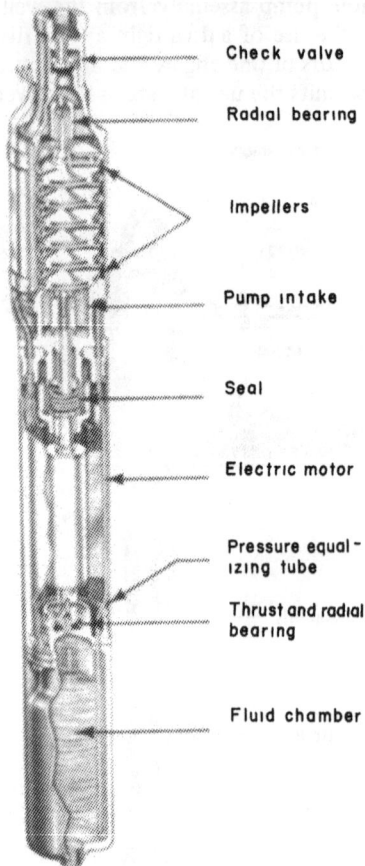

Check valve

Radial bearing

Impellers

Pump intake

Seal

Electric motor

Pressure equal-
izing tube

Thrust and radial
bearing

Fluid chamber

Fig. 8.14 CUT-AWAY VIEW OF A SUB-
MERSIBLE PUMP. (From F.
E. Myers & Bro. Company,
Ashland, Ohio.)

moving parts which are subject to both normal wear and that accelerated by corrosion and abrasive sand particles.

Submersible Pumps

Submersible pumps, though built during the past 50 years, have only been extensively used over the last 15 years. The increase in use coincided with design improvements in the submersible motors, electric cables and water-tight seals. These improvements made it possible to achieve efficiencies comparable with those obtained from lineshaft pumps and also long periods of trouble-free operation. The elimination of the long drive shaft (and multiple bearings with it) has not only eliminated the wearing and maintenance problems associated with lineshaft pumps but has also reduced the problems created by deviations in the vertical alignment of a well. The use of submersible pumps also results in savings in installation costs since pump houses are not usually required. The operation of the motor at a depth of several feet in the well also considerably reduces noise levels. The entire pump and motor must, however, be withdrawn to effect repairs and to service the motor. The need to do so, however, arises very infrequently.

PRIMING OF PUMPS

Priming is the name given to the process by which water is added to a pump in order to displace any air trapped in the pump and its suction pipe during shut-down periods. In other words, priming results in a continuous body of water from the inlet eye of the pump impeller downward through the suction pipe. Without this continuous body of water a centrifugal pump will not deliver water after the engine or motor has been started. Positive displacement types of pumps are less affected and need priming only to the extent necessary to seal leakage past pistons, valves and other working parts.

The many devices and procedures used in obtaining and maintaining a primed condition in pumps generally involve one or a combination of the

126

following: (1) a foot-valve to retain water in the pump during shut-down periods, (2) a vent to permit the escape of trapped air, (3) an auxiliary pump or other device (pipe from an overhead tank) to fill the pump with water, (4) use of a self-priming type of construction in the pump. Self-priming pumps usually have an auxiliary chamber integrated into the pump structure in such a way that the trapped air is exhausted as the pump circulates the priming water.

PUMP SELECTION

The proper selection of a pump for installation at a well involves the consideration of several factors. The following discussion presents some of the more important factors and particularly those which are very often overlooked and need to be emphasized.

The first factor to be considered must of necessity be the yield of the well. So logical as this may seem, it is a factor that is often overlooked in pump selection for small wells. There is no way of extracting more water from a well than that determined by its maximum yield. It is, therefore, foolhardy to select a pump of greater discharge capacity than the well will yield. Maximum well yields are usually determined by test pumping. For small wells, test pumping need not involve more than the pumping of the well at a specific rate or series of rates for a period of time in excess of the likely service requirements. The records of the test can then be used to determine the specific capacity.

With the knowledge of the specific capacity and the estimated water demands a suitable pumping rate can then be selected taking into consideration the provision of storage. Consideration may be given to the use of several hours of storage capacity and a high pumping rate in order to keep the number of pumping hours as low as possible. The advantages of so doing should be weighed against the use of a lower pumping rate for extended hours of pumping and the provision of lower storage capacity. The availability of electric power only for limited periods of the day or night would also influence the decision. Having chosen a pumping rate, the expected drawdown in the well for that rate can then be estimated by dividing it by the specific capacity of the well. For example, a pumping rate of 30 gpm in a well with a specific capacity of 5 gpm/ft would create a drawdown of 30 divided by 5, that is, 6 feet. Adding the drawdown to the depth of the static water level below the ground surface gives the depth to the expected pumping water level. This depth to the pumping water level is then used to choose between a surface-type pump and a deep well pump. In so doing, it must be remembered that seasonal variations in the water table, extended pumping and interference from other wells could cause the lowering of the pumping water level. Allowances should, therefore, be made for such possibilities where they are likely to occur. The use of deep well pumps would be indicated where the depth to the pumping water level is 25 feet or more and the well is deep enough and large enough in diameter to accommodate a suitable pump. Surface-type pumps would otherwise be used with limited pumping rates if necessary.

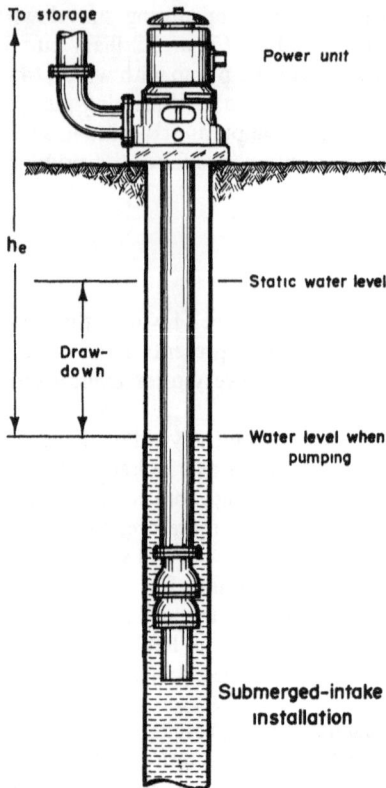

Fig. 8.15 TOTAL PUMPING HEAD OF WATER WELL PUMP INCLUDES VERTICAL LIFT, h_e, PLUS FRICTION LOSSES IN PIPE, h_f, AND VELOCITY HEAD (may usually be neglected).

The next logical step is the estimation of the total pumping head which, with the pumping rate, determines the capacity of the pump to be selected. The total pumping head, h_t, can be estimated by adding the total vertical lift, h_e, from the pumping water level to the point of delivery of the water (Fig. 8.15) and the total friction losses, h_f, occurring in the suction and delivery pipe. This estimate ignores the velocity head or head required to produce the flow through the system since this head can be expected to be negligible in most installations using small wells. The total vertical lift, h_e, includes the suction lift and the delivery head or head above the pump impeller when a surface-type pump is used (Fig. 8.2). The total friction losses, h_f, can be estimated with the use of Table B.10 in Appendix B.

Pump manufacturers or their agents can then be consulted on the selection of a suitable pump to meet the estimated pumping capacity and suction conditions, where applicable. A number of other factors would affect the final selection. Among these are the purchase price and cost of operating the pump; the extent of maintenance required and reliability of the maintenance service available; the availability of spare parts; the ease with which repairs can be effected; the sanitary features of the pump; and the desirability to standardize on the use of a particular type and make of pump in order to reduce the diversity of spare parts. A guide to pump selection is provided in Table 8.1. In it is summarized the conditions under which the various types of pumps discussed in this chapter would normally be used and the advantages and disadvantages of each type. It must be emphasized that the table is designed for use only as a general guide to pump selection.

SELECTION OF POWER SOURCE

The cost of power can and often does constitute a major part of the cost of pumping. In view of the limited economic resources usually available to

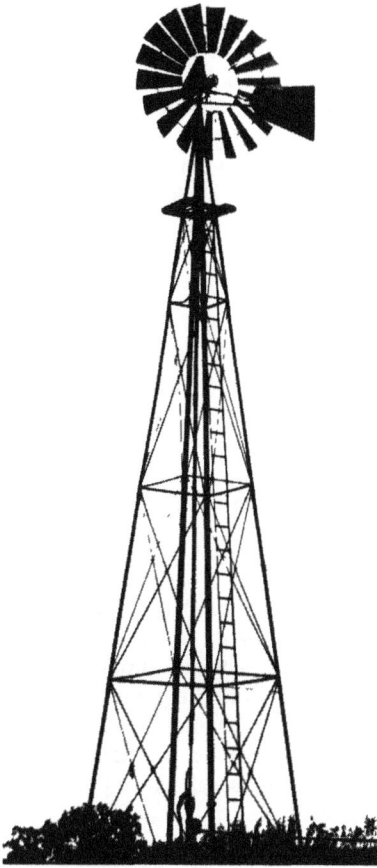

Fig. 8.16 W I N D M I L L . (Manual operation of pump also possible.)

those persons and communities using small wells, it is very important that careful consideration be given to the selection of a power source. The type of power available will, in many cases, be the determining factor in the design of a small pumping installation. There is normally a choice of four sources of power for operating pumps on small wells. These sources are man power, wind, electric motors and internal combustion engines.

Man Power

Man power is, in many places, not only a cheap source, but, sometimes, the only one available for operating pumps on wells. It is, of course, the oldest known source. Its use is suited to individual water supply systems with small, intermittent demands. Sometimes elevated storage is provided to maintain a continuous supply. The use of man power would, normally, be restricted to pumping rates not exceeding about 10 gpm and suction lifts of no more than about 20 feet. Hand pumps, subject to repeated use by the general public, can often have abnormal maintenance problems due to the fracturing of the hand lever and cylinder, and excessive wear of the inner wall of the cylinder, particularly when the water contains sand. The most sturdy types of pumps should be used under such conditions. Manufacturers have been experimenting with various types of metal construction of the levers, cylinders and cylinder linings in order to minimize the maintenance problems.

Wind

Wind is another very cheap source of power worthy of careful consideration in individual and small community water supply systems. Windmills (Fig. 8.16) usually require the availability of winds at sustained speeds of more than 5 miles per hour. Towers are normally used to raise the windmills 15 to 20 feet above the surrounding obstacles in order to provide a clear sweep of wind to the mills. Windmills usually drive reciprocating pumps through a connection of the pump rod from the mill to the piston rod of the pump.

Provision may also be made for pumping by hand during long periods of relative calm. It is good practice to provide adequate elevated storage to maintain the water supply during periods when there is insufficient wind. Windmills are normally manufactured in sizes expressed in terms of the diameters of their wheels. When ordering a windmill from a manufacturer, he must be supplied with information on the average wind velocity in addition to the required capacity and other relevant information on the pump. The operation and maintenance costs of windmills are usually very negligible and strongly influence their use in communities whose financial resources are inadequate to operate and maintain motor or engine driven pumps.

Electricity

Electricity, where available from a central supply at reasonable cost, is to be preferred over other sources of power. It would, however, be unwise to install electric generators simply to provide a supply for operating a small pump. Electricity's great advantage is the fact that it can be used to provide a continuous, automatically controlled supply of water. The power source must be reliable and not subject to significant voltage variations. Small electric motors are usually low in initial cost, require little maintenance and are cheap to operate.

Internal Combustion Engine

Internal combustion engines (gasoline, diesel or kerosene) are often used in areas where electric power is not available and winds are infrequent or inadequate to meet the water supply demands. Diesel engines, though usually the most costly to purchase, are generally the best from the point of view of operation and maintenance. Internal combustion engines require more maintenance than electric motors and must always be attended by an operator. Good service is obtained if a regular routine maintenance program is followed and a supply of spare parts always available.

TABLE 8.1–GUIDE TO PUMP SELECTION

(Adapted from Table 7-Information on Pumps. Manual of Individual Water Supply Systems, U.S. Dept. of Health, Education & Welfare, Public Health Service Pub. No. 24, Revised 1962)

Type of Pump	Practical Suction Lift* (ft)	Usual Pumping Depth (ft)	Usual Pressure Heads (ft of water)	Advantages	Disadvantages	Remarks
Reciprocating. 1. Surface 2. Deep Well	22-25 22-25	22-25 > 25	50-200 Up to 600 above the cylinder.	1. Positive action. 2. Discharge constant under variable heads. 3. Great flexibility in meeting variable demands. 4. Pumps water containing sand and silt. 5. Especially adapted to low capacity and high lifts.	1. Pulsating discharge. 2. Subject to vibration and noise. 3. Maintenance costs may be high. 4. May cause destructive pressure if operated against a closed valve.	1. Best suited for capacities of 5-25 gpm against moderate to high heads. 2. Adaptable to hand operation.
Rotary: 1. Surface (gear or vane)	22	22	50-250	1. Positive action. 2. Discharge constant under variable heads. 3. Efficient operation.	1. Subject to rapid wear if water contains sand or silt. 2. Wear of gears reduces efficiency.	1. Best suited for low speed operation. 2. Semi-rotary type adaptable to hand operation.
2. Deep Well (helical rotor)	Usually submerged.	> 25	100-500	1. Same as surface-type rotary. 2. Only one moving pump part in the well.	1. Subject to rapid wear if water contains sand or silt.	1. A cutless rubber stator increases life of pump. 2. Best suited for low capacity and high heads.

Type of Pump	Practical Suction Lift* (ft)	Usual Pumping Depth (ft)	Usual Pressure Heads (ft of water)	Advantages	Disadvantages	Remarks
Centrifugal: 1. Surface a. Volute (single stage)	20	10-20	100-150	1. Smooth, even flow. 2. Pumps water containing sand and silt. 3. Pressure on system is even and free from shock. 4. Low starting torque. 5. Usually reliable and good service.	1. Loses prime easily. 2. Good efficiency requires operation under design heads and speeds.	1. Better efficiencies at discharges near 50 gpm and heads up to about 150 ft.
b. Turbine (single stage)	28	28	100-200	1. Same as for volute type but not suitable for pumping water containing sand or silt. 2. They are self-priming.	1. Same as volute type except maintains priming easily.	1. Pressure reduction with increased capacity not as severe as for volute type.
2. Deep Well a. Vertical line-shaft turbine (multi-stage)	Impellers submerged.	> 25	100-800	1. Same as surface-type turbine.	1. Good efficiency requires operation under design heads and speeds. 2. Requires sufficiently straight and plumb well for installation and proper operation. 3. Lubrication and alignment of shaft critical. 4. Subject to abrasion from sand.	1. Severe maintenance problem when pumping corrosive water unless pump, column, shaft, etc. are made of non-corrosive materials.

Type of Pump	Practical Suction Lift* (ft)	Usual Pumping Depth (ft)	Usual Pressure Heads (ft of water)	Advantages	Disadvantages	Remarks
b. Submersible turbine (multi-stage)	Pump and motor submerged.	> 25	50-400	1. Same as surface-type turbine. 2. Short pump shaft to motor. 3. Plumbness and alignment of well less critical than for lineshaft type. 4. Less maintenance problems due to wearing of moving parts than for lineshaft type. 5. Lower installation and housing costs than for lineshaft type. 6. Lower noise levels during operation than for lineshaft type.	1. Repair to motor or pump requires removal from well. 2. Repair to motor may require shipment to manufacturer or his agent. 3. Subject to abrasion from sand.	1. Relatively recent design improvements for sealing of electrical equipment make long periods of trouble-free service possible. 2. Motor should be protected by suitable device against power failures.
Jet: 1. Deep Well	20-100 below the ground. (Ejector submerged 5 ft).	> 25	80-150	1. Simple in operation. 2. Does not have to be installed over the well. 3. No moving parts in the well. 4. Low purchase price and maintenance costs.	1. Generally inefficient. 2. Capacity reduces as lift increases. 3. Air in suction or return line will stop pumping.	1. The amount of water returned to the ejector increases with increased lift — 50% of total water pumped at 50 ft lift and 75% at 100 ft lift. 2. Generally limited to discharge of about 20 gpm against 150 ft maximum head.

*Practical suction lift at sea level. Reduce lift 1 foot for each 1000 ft above level.

133

CHAPTER 9

SANITARY PROTECTION
OF GROUND-WATER SUPPLIES

Too great stress can never be placed on the need to provide sanitary protection for all known ground-water sources, whether in immediate use or not, since such sources may some time in the future be of great importance to the development of their localities.

Small wells, of the type being considered in this manual, very often have relatively shallow aquifers as their sources of water. These sources, in many cases, are merely a few feet below the ground surface and can often be reached without great difficulty by pollution from privies, cesspools, septic tanks, barnyard manure, and industrial and agricultural waste disposal. It also very often happens that privies and septic tanks are the only economically practicable means of sewage disposal in a small and sparse community which must for various reasons depend entirely upon a shallow ground-water source for its potable water supply. Such dependence may be due to the inability of a small community to meet the costs of a sophisticated treatment plant for available surface water. Many rural areas are also subjected to annual extended periods of drought when streams become completely dry and shallow ground-water aquifers provide the only reliable sources of potable water. It is, therefore, of great importance that such sources be adequately protected.

POLLUTION TRAVEL IN SOILS

The sanitary protection of ground-water supplies must be based on an understanding of the basic facts relating to the travel of polluted substances through soils and water-bearing formations. It must be remembered that all water seeping into the ground is polluted to some degree, yet this water can later be retrieved in a completely satisfactory condition for domestic and other human uses. Some purifying processes must be taking place within the soil as the water travels through it. Several studies have been made of "nature's purifying action" by research workers in many parts of the world, particularly in Europe, India and the United States of America. These studies have contributed very much to our knowledge of the processes involved in the natural purification of ground waters and the patterns and extent of flow of pollution in them. The basic findings are summarized in the succeeding paragraphs.

The natural processes occurring in soils to purify water travelling through them are essentially three in number. The first two of these are the mechanical removal of microorganisms (including disease-producing bacteria) and other suspended matter by *filtration* and *sedimentation* or settling. Filtration depends upon the relative sizes of the pore spaces of the soil particles and those of the microorganisms and other filterable material. The finer the soil particles and the smaller the pore spaces between them, the more effective is the filtration process. Filtered material also tends to clog the pore spaces and thus help to improve the filtration process. Sedimentation depends upon the size of the suspended material and the rate of flow of the water through the pore spaces. The larger the particles of suspended matter and the slower the rate of flow through the soil, the more efficient would be the sedimentation process. It is, therefore, seen that the porosity and permeability of the soil are very important factors in the operation of the filtration and sedimentation processes and, as a result, in the extent of travel of bacterial pollution in soils.

The third factor is what is often termed the natural *die-away* of bacteria in soils. Bacteria which produce disease in man live for only limited periods of time outside of their natural host which is generally man or animals. Their life spans are usually short in the unfavorable conditions found in soils. This property contributes considerably to the self-purification of ground water during its movement and storage in sand and gravel aquifers.

The effect of filtration is, of course, completely lost and sedimentation somewhat reduced in ground waters travelling through large crevices and solution channels in limestone and other such consolidated rocks. This explains the generally better microbiological quality of ground water obtained from sands, gravels and other unconsolidated formations as against those obtained from the larger crevices, fissures and solution channels in consolidated rocks.

While the above mentioned processes are effective against the travel of bacterial pollution in ground water and usually within short distances and periods of time, they are not nearly as effective against the travel of chemical pollution. Chemical pollution, it will be seen later, persists much longer and travels much faster in ground waters than does bacterial pollution. Chemical reactions with soil material do play some part in restricting the travel of chemical pollution but require more time than the other processes do in controlling bacterial pollution.

Pollution (bacterial and chemical) in soils usually moves downward from the source until it reaches the water table and then along with the ground-water flow in a path which first gradually increases in width to a limited extent and then reduces to final disappearance. Downward travel of bacteria through homogeneous soil above the water table has seldom been found to be more than about 5 feet. Upon reaching the water table, no pollution travel takes place against the natural direction of ground-water flow unless induced by the pumping of a well upstream of the pollution source and with a circle of influence (upper surface of the cone of depression) that includes the pollution source. The horizontal path of the flow of bacterial pollution in sand formations from a point source, such as a well used to recharge an aquifer, has been found to reach a maximum width of about 6 feet before final disappearance at a distance of about 100 feet from the source. The corresponding distances for pit latrine sources have generally been smaller. The maximum distance of bacterial pollution flow is often reached several hours (often less than 2 days) after the introduction of the pollution. Filtration and the natural die-away processes then cause a rapid reduction in the numbers of bacteria found and the extent of the path until eventually only the immediate vicinity of the pollution source is found to be affected.

Chemical pollution follows a similar but much wider and longer path than that of bacterial pollution. Maximum widths of about 25 to 30 feet and lengths of about 300 feet have been observed. Investigations have indicated that chemical pollution travels twice as fast as bacterial pollution.

The above findings serve to emphasize the importance of the proper location of wells with respect to sources of pollution if contamination is to be avoided. They also form the basis for the general rules which are applied in well location and construction and the siting of pit latrines, cesspools and other such means of waste disposal in relation to ground-water sources.

WELL LOCATION

Wells should be located on the highest practicable sites and certainly on ground higher than nearby sources of pollution. The ground surface in the immediate vicinity of the well should slope away from it and be well drained. If necessary, the site should be built up to achieve this end. A special drainage system should be provided for waste water from public wells. It is good practice, whenever practicable, to off-set the pump installation and discharge pipe as far as possible from a public well. This, together with a good drainage system, ensures that no waste water accumulates in the immediate vicinity of the well to become a possible source of pollution or unsightly pools and mosquito breeding grounds. If a well must be located down-hill from a pollution source, then it should be placed at a reasonably safe distance away, depending upon the source and the soil conditions. Recommended minimum distances from various types of pollution sources are listed in Table 9.1.

TABLE 9.1

Pollution Source	Recommended Minimum Distance
Cast iron sewer with leaded or mechanical joints	10 ft.
Septic tank or sewer of tightly jointed tile	50 ft.
Earth-pit privy, seepage pit or drain field	75 ft.
Cesspool receiving raw sewage	100 ft.

These minimum distances are meant to be no more than guides to good practice and may be varied as soil and other conditions require. They should be applied only where the soil has filtering capacity equal to, or better than that of sand.

Well location should also take into consideration accessibility for pump repair, cleaning, treatment, testing and inspection. Wells located adjacent to buildings should be at least 2 feet clear of any projections such as over-hanging eaves.

Chapters 4, 5 and 6 should be consulted for information concerning the design, construction and completion aspects, respectively, of the sanitary protection of wells.

SEALING ABANDONED WELLS

The objectives of sealing abandoned wells are (1) the prevention of the contamination of the aquifer by the entry of poor quality water and other foreign substances through the well, (2) the conservation of the aquifer yield and artisean head where there is one, and (3) the elimination of physical hazard.

The basic concept of proper sealing of an abandoned well is the restoration, as far as practicable, of the existing geologic conditions. Under water-table conditions sealing must be effective to prevent the percolation of surface water through the well bore or along the outside of the casing to the water table. Sealing under artesian conditions must be effective in confining water to the aquifer in which it occurs.

Sealing is usually achieved by grouting with puddled clay, cement or concrete. When grouting under water, the grouting material should be placed from the bottom up by methods that would avoid segregation or excessive dilution of the material. Grouting methods have been described in Chapter 5.

It may be necessary, in some cases, to remove well casing opposite water-bearing zones to assure an effective seal. Where the upper 15 or 20 feet of the well casing was not carefully cemented during the original construction, this portion of the casing should be removed before final grouting for abandonment.

REFERENCES

Acme Fishing Tool Company. *Acme. Greatest Name in Cable Tools Since 1900.* Catalog. Parkersburg, West Virginia.

American Water Works Association, National Water Well Association. *A WWA Standard for Deep Wells.* AWWA A100-66. New York: AWWA Inc., 1966.

Anderson, Keith E. (ed.). *Water Well Handbook.* Rolla, Missouri: Missouri Water Well Drillers Association, 1966.

Baldwin, Helene L., and C. L. McGuinness. *A Primer on Ground Water.* United States Department of the Interior, Geological Survey. Washington: Government Printing Office, 1963.

Bowman, Isaiah. *Well-Drilling Methods.* Geological Survey Water-Supply Paper 257. Washington: Government Printing Office, 1911.

Decker, Merle G. *Cable Tool Fishing,* Water Well Journal. Series of articles commencing Vol. 21, No. 1. (Jan., 1967), pp. 14-16, 59.

Departments of the Army and the Air Force. *Wells.* TM5-297, AFM 85-23. Washington: Government Printing Office, 1957.

Edward E. Johnson, Inc. *Ground Water and Wells.* St. Paul, Minnesota, 1966.

Gordon, Raymond W. *Water Well Drilling with Cable Tools.* South Milwaukee, Wisconsin: Bucyrus-Erie Company, 1958.

Harr, M. E. *Groundwater and Seepage.* New York: McGraw-Hill Book Company Inc., 1962.

Livingston, Vern. *From: Too-Thin-to-Plough Missouri. To: Just-Right-to-Drink Well Water,* Water Works Engineering (May, 1957), pp. 493-495, 521.

McJunkin, Frederick E. (ed.). International Program in Sanitary Engineering Design. *Jetting Small Tubewells by Hand.* AID-UNC/IPSED Series Item No. 15. University of North Carolina, 1967.

Meinzer, O. E. *Occurrence of Ground Water in the United States.* Geological Survey Water-Supply Paper 489. Washington: Government Printing Office, 1959.

————. *Outline of Ground-Water Hydrology.* Geological Survey Water-Supply Paper 494. Washington: Government Printing Office, 1965.

Miller, Arthur P. *Water and Man's Health.* Washington: Department of State, Agency for International Development, 1962.

New York State Department of Health. *Rural Water Supply.* Albany, New York, 1966.

State of Illinois, Department of Public Health, Division of Sanitary Engineering. *Illinois Water Well Construction Code.* Springfield, Illinois, 1967.

Todd, David Keith. *Ground Water Hydrology.* New York: John Wiley & Sons, Inc., 1960.

U. S. Department of Health, Education, and Welfare. *Drinking Water Standards.* Public Health Service Publication No. 956. Washington: Government Printing Office, 1962.

_____. *Manual of Individual Water Supply Systems.* Public Health Service Publication No. 24. Washington: Government Printing Office, 1963.

U. S. Department of the Interior, Interdepartmental Committee on Water for Peace. *Water for Peace. A Report of Background Considerations and Recommendations on the Water for Peace Program.* Washington: Government Printing Office, 1967.

Wagner, E. G., and J. N. Lanoix. *Water Supply for Rural Areas and Small Communities.* Geneva: World Health Organization, 1959.

Wisconsin State Board of Health. *Wisconsin Well Construction and Pump Installation Code.* Madison, Wisconsin, 1951.

CREDIT FOR ILLUSTRATIONS

The authors wish to give credit to UOP-Johnson Division, Universal Oil Products Company, St. Paul, Minnesota for the use and adaptation of the following illustrations: Figures 2.3, 2.5, 2.9, 2.10, 2.12, 2.13, 2.17, 3.1, 3.2, 4.1, 4.2, 4.3, 4.5, 4.7, 4.8, 4.10, 4.11, 5.16, 5.22, 5.23, 5.24, 5.25, 5.26, 5.29, 5.30, 5.31, 5.32, 5.33, 5.34, 5.35, 5.36, 5.37, 5.38, 5.39, 5.40, 5.41, 6.1, 6.2, 6.4, 6.5, 6.6, 6.8, 7.1, 7.2, 8.2, and A.1.

APPENDIX A

MEASUREMENT OF PERMEABILITY

The coefficient of permeability, or permeability as it is usually referred to in practice, can be determined by both laboratory and field experiments. The field experiments, or pumping tests as they are called, have the advantage over laboratory experiments in that they are performed on the aquifer materials in their natural, undisturbed state. They are, however, more complicated, time consuming, costly and beyond the scope of this book.

Permeameters are used for laboratory determinations of permeability. The simplest laboratory method for the determination of permeability uses a constant head permeameter as follows. Flow under constant head or pressure is maintained through a chosen length, ℓ, of the sample of aquifer material placed between porous plates in a tube of cross sectional area, A (Fig. A.1). The device at the upper left of the figure is used to provide the flow under constant head. The rate of flow, Q, through the sample is obtained by measuring the volume, V, of water discharged into a graduated cylinder in a given time, t. The manometer tubes to the right of the figure are used to measure the head loss, $h_1 - h_2$, as the water flows through the length, ℓ, of the sample. Care must be taken to expel any air trapped in the sample before taking measurements.

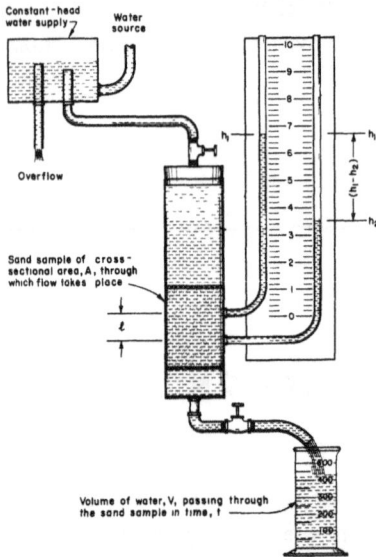

Fig. A.1 CONSTANT-HEAD PERMEA-METER FOR LABORATORY DETERMINATION OF COEFFECIENTS OF PERMEABILITY.

Then
$$Q = \frac{V}{t} = \frac{P(h_1 - h_2)A}{\ell}$$

Giving
$$P = \frac{V\ell}{(h_1 - h_2)\,A\,t}$$

To obtain P in units of gallons per day per square foot (gpd/sq ft), V must be expressed in gallons, ℓ, h_1, and h_2 in feet, A in square feet, and t in days.

140

APPENDIX B

USEFUL TABLES AND FORMULAS

TABLE B.1 LENGTH

Unit	Equivalents of First Column						
	Centi-meters	Meters	Kilo-meters	Inches	Feet	Yards	Miles
1 Centimeter	1	.01	.00001	.3937	.0328	.0109	.0000062
1 Meter	100	1	.001	39.37	3.2808	1.0936	.000621
1 Kilometer	100,000	1,000	1	39,370	3,280.8	1,093.6	.621
1 Inch	2.54	.0254	.0000254	1	.0833	.0278	.000016
1 Foot	30.48	.3048	.000305	12	1	.3333	.000189
1 Yard	91.44	.9144	.000914	36	3	1	.000568
1 Mile	160,935	1,609.3	1.6093	63,360	5,280	1,760	1

TABLE B.2 AREA

Unit	Equivalents of First Column						
	Square Centimeters	Square Meters	Square Inches	Square Feet	Square Yards	Acres	Square Miles
1 Sq. centimeter	1	.0001	.155	.00108	.00012	–	–
1 Sq. Meter	10,000	1	1,550	10.76	1.196	.000247	–
1 Sq. Inch	6 452	.000645	1	.00694	.000772	–	–
1 Sq. Foot	929	.0929	144	1	.111	.000023	–
1 Sq. Yard	8,361	.836	1,296	9	1	.000207	–
1 Acre	40,465,284	4,047	6,272,640	43,560	4,840	1	.00156
1 Sq. Mile	–	2,589,998	–	27,878,400	3,097,600	640	1

TABLE B.3 VOLUME

Unit	Equivalents of First Column						
	Cubic Centimeters	Cubic Meters	Liters	U.S. Gallons	Imperial Gallons	Cubic Inches	Cubic Feet
1 Cu. Centimeter	1	.000001	.001	.000264	00022	.061	.0000353
1 Cu. Meter	1,000,000	1	1,000	264.17	220.083	61,023	35.314
1 Liter	1,000	.001	1	.264	.220	61,023	.0353
1 U.S Gallon	3,785.4	.00379	3.785	1	.833	231	.134
1 Imperial Gallon	4,542.5	.00454	4.542	1.2	1	277.274	.160
1 Cu. Inch	16.39	.0000164	.0164	.00433	.00361	1	.000579
1 Cu. Foot	28,317	.0283	28.317	7.48	6.232	1,728	1

TABLE B.4 FLOW

Unit	Equivalents of First Column						
	Cubic Feet Per Second	Cubic Feet Per Day	U.S. Gallons Per Minute	Imp. Gallons Per Minute	U.S. Gallons Per Day	Imp. Gallons Per Day	Acre Feet Per Day
1 Cu. Foot per Sec.	1	86,400	448.83	374.03	646,323	538,860	1.983
1 Cu. Foot per Day	.0000116	1	.00519	.00433	7.48	6.233	.000023
1 U.S. Gallon per Min.	.00223	192.50	1	.833	1,440	1,200	.00442
1 Imp. Gallon per Min.	.00267	231.12	1.2	1	1,728	1,440	.0053
1 U.S. Gallon per Day	.00000155	.134	.000694	.000579	1	.833	.00000307
1 Imp. Gallon per Day	.00000186	.160	.000833	.000694	1.2	1	.0000368
1 Acre Foot per Day	.504	43,560	226.28	188.57	325,850	271,542	1

TABLE B.5 WEIGHT

Unit	Equivalents of First Column					
	Grams	Kilograms	Ounces (Avoirdupois)	Pounds (Avoirdupois)	Tons (Short)	Tons (Long)
1 Gram	1	.001	.0353	.0022	.0000011	.00000098
1 Kilogram	1000	1	35 274	2.205	.0011	.000984
1 Ounce (Avoirdupois)	28.349	.0283	1	.0625	.0000312	.0000279
1 Pound (Avoirdupois)	453.592	.454	16	1	.0005	.000446
1 Ton (Short)	907,184.8	907.185	32,000	2,000	1	893
1 Ton (Long)	1,016,046.98	1,016.047	35,840	2,240	1.12	1

TABLE B.6 POWER

Unit	Equivalents of First Column				
	Watts	Kilowatts	Horsepower	Foot Pounds Per Minute	Joules Per Second
1 Watt	1	.001	.00134	44.254	1
1 Kilowatt	1000	1	1.341	44,254	1,000
1 Horsepower	746	.746	1	33,000	746
1 Foot Pound Per Minute	.0226	.0000226	.0000303	1	.0226
1 Joule Per Second	1	.001	.00134	44.254	1

TABLE B.7 VOLUMES AND WEIGHT EQUIVALENTS (Water at 39.2°F)

Unit	Equivalents of First Column						
	Cubic Meters	Liters	U.S. Gallons	Imp. Gallons	Cubic Inches	Cubic Feet	Pounds
1 Cu. Meter	1	1,000	264.17	220.083	61,023	35.314	2,200.83
1 Liter	.001	1	.264	.220	61.023	.0353	2.201
1 U.S. Gallon	.00379	3.785	1	.833	231	.134	8 333
1 Imp. Gallon	.00454	4.542	1.2	1	277.274	.160	10
1 Cu. Inch	.0000164	.0164	.00433	.00361	1	.000579	.0361
1 Cu. Foot	.0283	28.317	7.48	6.232	1,728	1	62.32
1 Pound	.00045	.454	.12	.1	27.72	.016	1

B.8 PRESSURE

1 Atmosphere = 760 millimeters of mercury at 32°F.
29.921 inches of mercury at 32°F.
14.7 pounds per square inch.
2,116 pounds per square foot.
1.033 kilograms per square centimeter.
33.947 feet of water at 62°F.

B.9 TEMPERATURE

Degrees C = 5/9 x (F − 32)   Degrees F = 9/5 C + 32

TABLE B.10   FRICTION LOSS IN SMOOTH PIPE
(approximate head loss in feet per 1000 feet of pipe)

Flow Rate in Gallons per Minute	Nominal Pipe Size in Inches					
	1¼	1½	2	2½	3	4
10	20	9	2			
15	44	20	6			
20	79	35	10	4	1	
25	123	55	16	6	2	
30	178	79	22	9	3	
40		142	40	16	5	
50		222	64	25	8	2

TABLE B.11   PIPE, CYLINDER OR HOLE CAPACITY

Diameter (Inches)	Gallons Per Foot
1½	0.09
2	0.16
2½	0.25
3	0.37
4	0.67
6	1.47
8	2.61
10	4.08
12	5.86
16	10.45
18	13.20
20	16.35
24	23.42

B.12 DISCHARGE MEASUREMENT USING SMALL CONTAINER

(Oil Drums, Stock Tanks, etc.)

$$\text{Discharge (Gallons per minute)} = \frac{\text{Volume of container (Gallons) x 60}}{\text{Time (Seconds) to fill container}}$$

TABLE B.13 ESTIMATING DISCHARGE FROM A HORIZONTAL PIPE FLOWING FULL

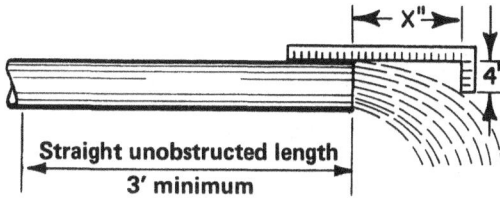

DISCHARGE RATE (Gallons Per Minute)

Horizontal Distance, x (Inches)	Nominal Pipe Diameter (Inches)				
	1	1¼	1½	2	2½
4	5.7	9.8	13.3	22.0	31.3
5	7.1	12.2	16.6	27.5	39.0
6	8.5	14.7	20.0	33.0	47.0
7	10.0	17.1	23.2	38.5	55.0
8	11 3	19.6	26.5	44.0	62.5
9	12.8	22.0	29.8	49.5	70.0
10	14.2	24.5	33.2	55.5	78.2
11	15.6	27.0	36.5	60.5	86.0
12	17.0	29.0	40.0	66.0	94.0
13	18.5	31.5	43.0	71.5	102.0
14	20.0	34.0	46.5	77.0	109.0
15	21.3	36.3	50.0	82.5	117.0
16	22.7	39.0	53.0	88.0	125.0

TABLE B.14 ESTIMATING DISCHARGE FROM VERTICAL PIPE OR CASING

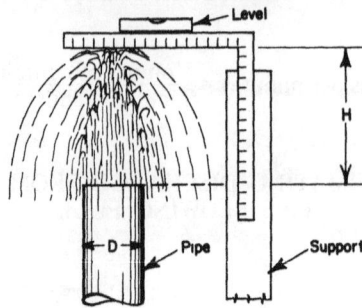

DISCHARGE RATE (Gallons Per Minute)

Height, H (Inches)	Nominal Pipe Diameter, D (Inches)		
	2	3	4
1½	22	43	68
2	26	55	93
3	33	74	130
4	38	88	155
5	44	99	175
6	48	110	190
8	56	125	225
10	62	140	255
12	69	160	280
15	78	175	315
18	85	195	350
21	93	210	380
24	100	230	400

146

TABLE B.15 ROPE CAPACITY OF DRUM OR REEL

(Wire Rope Evenly Spooled)

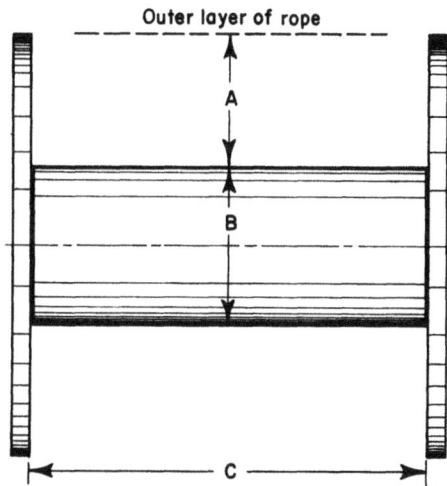

Outer layer of rope

Rope Capacity (feet) = K (A+B)AxC   where A,B,C are in inches and K has values in table below

Nominal Rope Diameter (Inches)	K	Nominal Rope Diameter (Inches)	K
1/4	4.4	9/16	.9
5/16	2.8	5/8	.7
3/8	2.0	3/4	.5
7/16	1.4	7/8	.4
1/2	1.1	1	.3

TABLE B.16 DRILLING CABLE ROPE CAPACITIES

(Left Laid — Mild Plow Steel — 6x19 Hemp Center)

Rope Diameter (Inches)	Approximate Weight Per Foot (Pounds)	Recommended Working Load (Pounds)
1/2	.42	3,200
9/16	.53	4,200
5/8	.66	5,000
3/4	.95	7,200
7/8	1.29	9,800
1	1.68	12,600

TABLE B.17 SAND LINE CABLE ROPE CAPACITIES

(Coarse Laid — Plow Steel — 6x7 Hemp Center)

Rope Diameter (Inches)	Approximate Weight Per Foot (Pounds)	Recommended Working Load (Pounds)
1/4	.09	800
5/16	.15	1,200
3/8	.21	1,800
7/16	.29	2,400
1/2	.38	3,200

TABLE B.18 CASING LINE CABLE ROPE CAPACITIES

(Non Rotating – Plow Steel – 18 x 7 Hemp Center)

Rope Diameter (Inches)	Approximate Weight Per Foot (Pounds)	Recommended Working Load (Pounds)
5/8	.68	5,400
3/4	.97	7,600
7/8	1.32	10,200

TABLE B.19 MANILA ROPE CAPACITIES (3-STRAND)

Rope Diameter (Inches)	Approximate Weight Per Foot (Pounds)	Recommended Working Load (Pounds)
3/8	.04	270
7/16	.05	350
1/2	.08	530
9/16	.10	690
5/8	.13	880
3/4	.17	1,080
7/8	.23	1,540
1	.27	1,800

149

INDEX

P

Particle classification, 43
— size analysis, (see Sieve analysis)
Particles, arrangement of, 14f.
—, distribution of, 14f., 46
—, packing of, 7, 14f.
—, rounding of, 7
—, shape of, 9
—, size of, 9, 14f.
—, sorting of, 7
Percent size, definition, 43
Percolation, benefits of, 3, 23
Percussion drilling, (see Drilling Methods)
Permeability, 12ff., 31, 43, 100, 135
—, coefficient of, 12ff.
—, factors affecting, 14f.
Permeable, 7f.
Pesticides, 27
pH, 26
Piezometric surface, definition, 10, 13, 16
Pipe, black iron, 109f.
— clamp, 70, 85
— joints, spigoted, 39
—, plastic, 38f., 49f., 109f.
—, polyvinyl chloride (pvc), 49
—, slotted, 33, 38f., 102
Pit latrine, 136
Plains, coastal, 31
—, river, 31
Pollution, bacterial, 23, 135f.
—, chemical, 135f.
—, rate of travel, 136
—, sources of, 23, 134ff.
— travel in soils, 134ff.
Polyphosphates, 104, 111
Pore, 4ff., 11, 135
Pores, continuity of, 14f.
—, volume of, 14
Porosity, 14f., 135
—, definition, 11
Porous, 7f.
Power consumption, 107f.
Precipitation, 4, 18
Pressure, 12f., 16f., 26, 40, 101, 107, 114ff., 121
— aquifer, 10
—, atmospheric, 115
— differences, 13
Privies, 23, 25, 134
Public health, 1
Pulling ring, 77, 85
Pump, 10, 114ff.
—, capacity of, 127f., 131ff.
—, centrifugal, 120ff., 132f.
—, —, turbine, 121f., 132
—, —, volute, 121, 132
—, constant displacement, 117ff., 131
—, — —, helical rotor, 117, 119f., 124, 131
—, — —, reciprocating piston, 117, 124, 129f.
—, — —, — —, discharge rate, 119
—, — —, rotary, 117, 119, 131
—, deep well, 116, 118, 120ff., 127, 132f.
—, — —, lineshaft, 124ff., 132
—, — —, submersible, 74, 124, 126, 133
—, driving force, 114, 117
—, hand-driven, 114, 119, 131, (also see manually-operated)
— house, 125

Pump impeller, 121f., 126, 128
—, jet, 120, 122ff., 133
—, lineshaft., 124ff., 132
—, manually-operated, 114, 118
—, multi-stage, 122ff.
—, pitcher, 118
—, positive displacement, 124, 126, 131, (also see constant displacement)
—, power source, electricity, 127, 129f.
—, — —, internal combustion engine, 129f.
—, — —, man power, 129
—, — —, selection of, 128ff.
—, — —, wind, 129f.
—, priming of, 126f.
—, reduction in capacity, 106
—, selection of, 127f.
—, self-priming, 127
—, standardization, 128
—, submersible, 74, 124, 126, 133
—, surface-type, 116, 118ff., 127f., 131f.
—, variable displacement, 117, 120ff., 132f.
—, — —, centrifugal, 120ff., 132f.
—, — —, jet, 120, 122ff., 133
—, vertical turbine, 74
Pumping equipment, 33f., 114ff.
—, hours of operation, record keeping, 107
— water level, 54, 116, 127
—, rates of, 107, 109, 119f., 124, 131ff.
— test, 127

Q

Quality, ground water, 3f., 20, 22ff., 31, 54
—, — —, chemical, 23ff., 48f.
—, — —, compared with surface water, 22f.
—, — —, microbiological, 22f., 135
—, — —, physical, 22

R

Radius of influence, 16
Recharge, 17f., 20
— area, 11, 18, 29
— —, definition, 11
— effects, 17f.
River beds, 32
Rocks, classification, 7ff.
—, consolidated, 7, 22, 29, 135
—, definition, 7
—, deposition of, 7
—, erosion of, 7
—, extrusive, 8
—, hard, 7
—, igneous, 7ff., 24
—, intrusive, 8
—, metamorphic, 7, 9
—, plutonic, 8
—, sedimentary, 7ff.
—, soft, 7
—, transport of, 7
—, unconsolidated, 1, 7, 29
—, volcanic, 8f.
—, weathering of, 7
Rope socket, 67, 93, (also see Wire line socket)
Rotary drilling, (see Drilling methods)